Get the eBook FREE!

(PDF, ePub, Kindle, and liveBook all included)

We believe that once you buy a book from us, you should be able to read it in any format we have available. To get electronic versions of this book at no additional cost to you, purchase and then register this book at the Manning website.

Go to https://www.manning.com/freebook and follow the instructions to complete your pBook registration.

That's it!
Thanks from Manning!

Idiomatic Rust

CODE LIKE A RUSTACEAN

BRENDEN MATTHEWS

MANNING
SHELTER ISLAND

For online information and ordering of this and other Manning books, please visit
www.manning.com. The publisher offers discounts on this book when ordered in quantity.
For more information, please contact

Special Sales Department
Manning Publications Co.
20 Baldwin Road
PO Box 761
Shelter Island, NY 11964
Email: orders@manning.com

Manning Publications Co.
20 Baldwin Road
PO Box 761
Shelter Island, NY 11964

Development editor: Karen Miller
Technical editor: Alain Couniot
Review editor: Kishor Rit
Production editor: Kathy Rossland
Copy editor: Keir Simpson
Proofreader: Melody Dolab
Technical proofreader: Jerry Kuch
Typesetter: Dennis Dalinnik
Cover designer: Marija Tudor

ISBN: 9781633437463
Printed in the United States of America

To my best friends, Doge and Walter,
without whom this book would not have been possible.
Thanks to them for their unrelenting optimism and support.

brief contents

contents

preface

When I started learning programming in the 1990s, I didn't have access to the sort of resources that are easy to find today. I didn't have proper internet access, and the World Wide Web was still in its infancy, so I relied primarily on trial and error. The library at my junior high school was (sadly) not stocked with computer science books.

It wasn't until many years later that I had access to learning resources such as books. At that point, I had already learned quite a bit merely by reading source code, experimenting, and asking questions on Internet Relay Chat (IRC) and forums. My teachers were mostly kind strangers on the internet, and I am grateful for their help.

Luckily, learning programming has never been easier, as innumerable high-quality resources are available today. In writing this book, I wanted to create something that I would have found helpful while I was learning programming. I hope that this book will benefit you and help you become a better programmer or achieve your goals, much as those kind strangers on the internet did for me so many years ago.

acknowledgments

I want to thank my good friends Javeed Shaikh and Ben Lin for their feedback on the early drafts of this book and for helping me work through various ideas. I'd also like to thank Eleanor Seay for her inspiration and support. Ava and Tobias, thank you for your patience and understanding.

I thank Manning Publications and its staff for all the support and help they gave me. Many thanks to development editor Karen Miller, technical proofreader Jerry Kuch, and all the members of the production team.

Special thanks to technical editor Alain Couniot. Alain is a long-time IT professional with a keen interest in innovation and programming languages—in particular, functional ones. His interests range from embedded systems to distributed enterprise applications, cloud and high-performance computing to quantum computing. Rust is currently his favorite language.

Thanks to all the reviewers—Alessandro Campeis, Andy Stainer, Charles Chan, David Paccoud, David White, Eder Andrés Ávila Niño, Filip Mechant, Florian Braun, Geert Van Laethem, George Reilly, Giuseppe Catalano, Guillaume Schmid, John Guthrie, Jon Christiansen, Lev Veyde, Martin Nowack, Scott Ling, Sergio Britos, Seung-jin Kim, Stefaan Verscheure, Stephen Wakely, Thomas Lockney, Volker Roth, Walter Alexander Mata López, William Wheeler, and Yves Dorfsman. Your suggestions helped make this a better book.

The patterns presented in this book are derived mainly from other people's great work, which I credit where due. In writing this book, I stood on the shoulders of giants, predominantly random people on the internet who have a passion for writing great software. I'm grateful and humbled that so many bright people are building beautiful things and sharing them with the world.

about this book

This book is a collection of design patterns and best practices for the Rust programming language, written to appeal to a broad audience of Rust programmers, from beginners to advanced developers. Some parts of this book take a theoretical approach, but most of them focus on practical use. My aim is to help you become a better Rust programmer by teaching you how to write idiomatic Rust code and use Rust's features effectively.

This book grew out of my other book, *Code Like a Pro in Rust* (Manning Publications, 2024), which is a more general guide to practical Rust and may be an excellent place for beginners to start before reading this book.

The original working title of this book was *Rust Design Patterns,* and the inspiration for it came from the classic *Design Patterns: Elements of Reusable Object-Oriented Software* (Addison-Wesley Professional, 1994). Although this book doesn't translate the original book's patterns directly into Rust, it's a collection of Rust-specific patterns and practices inspired by the original Design Patterns. It also became clear that the book was about more than design patterns, so the title was changed to *Idiomatic Rust: Code like a Rustacean* to better reflect the book's content.

How is this book different?

This book isn't intended to be a comprehensive guide to Rust or a reference manual for syntax or standard library functions. The patterns and practices presented in this book are designed to help you write better Rust and give you a deeper understanding of Rust and how to use it effectively.

Much of the discussion in this book focuses on patterns and practices that are not necessarily described or documented in the official Rust documentation and resources. However, you'll still find these patterns in use in many Rust codebases. Although these patterns are not always unique to Rust, they are presented here in the context of Rust programming.

Who should read this book?

This book is for Rust programmers at all skill levels, but beginning Rust programmers may find some of the content challenging. The book is not a beginner's guide to Rust, and it assumes that you have some familiarity with the Rust programming language.

Readers will benefit greatly from being familiar with the classic *Design Patterns: Elements of Reusable Object-Oriented Software*, as this book references the original design patterns and practices described in that book.

How this book is organized

This book is organized into four parts, each consisting of chapters that cover a specific aspect of Rust programming.

Part 1 is a review of Rust's core features and building blocks:

- Chapter 1 discusses the content of the book and introduces design patterns.
- Chapter 2 presents the basic building blocks of Rust.
- Chapter 3 reviews pattern matching and functional programming.

Part 2 goes into detail on Rust's core patterns and library design:

- Chapter 4 introduces core patterns in Rust.
- Chapter 5 presents Rust design patterns.
- Chapter 6 discusses library design.

Part 3 covers more advanced patterns in Rust:

- Chapter 7 discusses advanced techniques and patterns in Rust.
- Chapter 8 builds on the topics in chapter 7.

Part 4 discusses how to avoid problems and build robust software:

- Chapter 9 discusses immutability and how it's used in Rust.
- Chapter 10 presents several antipatterns and shows how to avoid them.

How to read this book

You can read this book from start to finish or jump around to the chapters that interest you most. Each chapter is designed to be self-contained so that you can read in any order, but some chapters reference concepts or patterns from earlier chapters. For less-experienced Rust programmers, reading the book in order may be helpful, as the patterns build on one another.

I recommend reading the book with a computer nearby so you can try the code samples and experiment with the patterns and practices described in the book. The best way to learn programming is to do it, so I encourage you to experiment with the code samples and apply the patterns and practices to your projects. The code samples are liberally licensed, so you can reuse them in your projects.

As described in Mortimer J. Adler's *How to Read a Book*, (Touchstone, 1974), you may get the most out of this book by reading it multiple times. The first time, you might focus on understanding the patterns and practices it presents. In subsequent readings, focus on applying the patterns and practices to your projects and experimenting with the code samples.

About the code

This book contains numerous original code samples. To obtain a copy of the source code, you can clone the book's Git repository on your local machine, hosted on GitHub at https://github.com/brndnmtthws/idiomatic-rust-book. The code samples presented are often partial, so you'll need to refer to the source code for the complete code listings.

The source code in the book's text may differ slightly from the code in the book's repository due to formatting and other presentation-related considerations, including line wrapping, indentation, and compilation (intentional errors shown in the book).

Examples of source code are in both numbered listings and inline with normal text. In both cases, source code is formatted in a `fixed-width font like this` to separate it from ordinary text. Sometimes, code is also in `bold` to highlight code that has changed from previous steps in the chapter, such as when a new feature adds to an existing line of code.

In many cases, the original source code has been reformatted; line breaks and reworked indentation have been added to accommodate the available page space in the book. In rare cases, even this change was not enough, and listings include line-continuation markers (➡). Additionally, comments in the source code may be removed from the listings when the code is described in the text. Code annotations accompany many of the listings, highlighting important concepts.

Over time, the code samples in the book may become outdated as the Rust language and ecosystem evolve. The code in the book's repository, however, will be updated to reflect the latest changes. I recommend referring to the book's repository for the most up-to-date code samples.

You can clone a copy of the book's code locally on your computer by running the following command in Git:

```
$ git clone https://github.com/brndnmtthws/idiomatic-rust-book
```

The book's code is organized in directories by chapter and section within the repository, which is itself organized within each section by topic. The code is licensed under the Massachusetts Institute of Technology (MIT) license, a permissive license that allows you to copy the code samples and use them as you see fit, even as the basis for your own work.

You can get executable snippets of code from the liveBook (online) version of this book at https://livebook.manning.com/book/idiomatic-rust. The complete code for the examples in the book is available for download from the Manning website at https://www.manning.com/books/idiomatic-rust and from GitHub at https://github .com/brndnmtthws/idiomatic-rust-book.

liveBook discussion forum

Purchase of *Idiomatic Rust: Code like a Rustacean* includes free access to liveBook, Manning's online reading platform. Using liveBook's exclusive discussion features, you can attach comments to the book globally or to specific sections or paragraphs. It's a snap to make notes for yourself, ask and answer technical questions, and receive help from the author and other users. To access the forum, go to https://livebook.manning .com/book/idiomatic-rust/discussion. You can also learn more about Manning's forums and the rules of conduct at https://livebook.manning.com/discussion.

Manning's commitment to our readers is to provide a venue where meaningful dialogue between individual readers and between readers and the author can take place. It is not a commitment to any specific amount of participation on the part of the author, whose contribution to the forum remains voluntary (and unpaid). We suggest that you try asking the author some challenging questions lest his interest stray! The forum and the archives of previous discussions will be accessible on the publisher's website as long as the book is in print.

about the author

BRENDEN MATTHEWS is a software engineer, entrepreneur, and prolific open source contributor. He has used Rust since the early days of the language and has contributed to several Rust tools and open source projects in addition to using Rust professionally. He's the author of Conky, a popular system monitor, and a member of the Apache Software Foundation with more than 25 years of industry experience. Brenden is also a YouTube contributor and instructor, as well as a writer of many articles on Rust and other programming languages. He has given talks at a number of technology conferences, including Qcon, Linux-Con, ContainerCon, MesosCon, and All Things Open, as well as at Rust meetups. He has been a GitHub contributor for more than 13 years, with multiple published Rust crates. He has contributed to several open source Rust projects and built production-grade Rust applications professionally.

about the cover illustration

The figure on the cover of *Idiomatic Rust: Code like a Rustacean*, titled "L'agent de change" ("The stockbroker"), is taken from a book by Louis Curmer published in 1841. The illustration is finely drawn and colored by hand.

In those days, it was easy to identify where people lived and what their trade or station in life was by their dress alone. Manning celebrates the inventiveness and initiative of the computer business with book covers based on the rich diversity of regional culture centuries ago, brought back to life by pictures from collections such as this one.

Part 1

Building blocks

We'll begin this book by examining some of the basic building blocks of Rust design patterns. These building blocks are essential to understanding the complex patterns we'll cover later in the book, and they'll help you write more idiomatic Rust code. Some of these building blocks are specific to Rust; others are more general programming concepts that are particularly important in Rust.

These building blocks are effectively the vocabulary of *Idiomatic Rust: Code like a Rustacean* and constitute the core features of the Rust language. We can think of them as the atoms of a molecule, which we'll combine in various ways to create complex substances (or patterns). Those patterns can be combined and architected to create an endless variety of software systems.

Building on a solid foundation allows us to achieve great heights, provided that we build a solid and sound structure with care and attention. Rust provides an excellent foundation, but ultimately, we developers are responsible for deciding how to use the tools and components at our disposal effectively.

Rust-y patterns

Reading this book is a great way to advance your Rust skills, whether you're a beginner, intermediate, or advanced Rust programmer. If you're a beginner, studying design patterns is an excellent path to elevate your skills above the basics of the Rust language, but you may find some parts of this book challenging, so you may need to study other resources as you go. This book presents a variety of techniques for writing high-quality Rust code, but we'll focus on patterns, idioms, and conventions that are widely used and understood by the Rust community.

Design patterns are powerful abstractions that every programmer can use to produce high-quality code. Humans are excellent at pattern recognition, and following well-understood and easily recognized patterns helps us solve two tricky problems: reasoning about whether a design is good or bad (following well-known patterns helps us avoid creating bad code, for example) and helping other people understand our code.

Reading code is often more challenging than writing code. When we read someone else's code that follows well-understood patterns, it's easier to reason about what the code is doing if we recognize the patterns. If you've trained your brain to recognize the most common patterns, judging code quality becomes much more manageable, resulting in fewer mistakes. We can take advantage of millions of years of evolution by teaching our brains which patterns to recognize, short-circuiting the challenge of judging code quality.

When it comes to writing code, knowing which patterns to apply in which situations helps us produce good code in less time. This knowledge is no different from learning which data structures or algorithms to use in other circumstances and the trade-offs that come with them.

You won't find much dogma in this book. I'll do my best to present the patterns along with detailed explanations of why we're doing what we do. You, as a programmer, are free to experiment, diverging from the patterns presented in this book to create your own designs. I'll offer opinionated conventions, however, generally preferring convention to configuration.

To use an analogy, I prefer going to a restaurant where the chef offers one or two items on the menu, preselected as the best items for the season, to scanning a menu of tens or even hundreds of dishes and trying to figure out which are best. The best restaurants generally provide a curated experience (you trust the chef's good taste), and I hope to do the same with this book.

Many of the code samples in this book are partial listings, but you can find the full working code samples on GitHub at https://github.com/brndnmtthws/rust-advanced -techniques-book. The code is available under the Massachusetts Institute of Technology (MIT) license, which permits use, copying, and modifications without restriction. If you can, I recommend that you follow along with the full code listings to get the most out of this book. The code samples are organized by chapter within the repository; some examples, however, span multiple sections or chapters and are named based on their subject matter. The code samples in the book may differ slightly from those in the repository, as the book's code samples are edited for clarity, brevity, and suitability for print.

1.1 *What this book covers*

In this book, I'll present various idioms, patterns, and design patterns. Some of these patterns are specific to Rust; others are old ideas presented in a new format within the framework of Rust's unique features, grammar, and syntax.

This book aims to help you understand and apply these patterns to improve your software design and architecture. Learning and using these patterns allows you to write more efficient, maintainable, scalable code. Throughout the book, I'll explain each pattern, including why it is important and how it can be applied in real-world scenarios. I'll also discuss the tradeoffs and considerations involved in using each pattern.

It's worth noting that you should not follow design patterns blindly. Patterns are tools that you can adapt and modify to suit your specific needs. As a programmer, you have the freedom to experiment and diverge from the patterns presented in this book to create your own unique designs. By the end of this book, you'll have a solid understanding of various idioms, patterns, and design patterns in Rust, and you'll be equipped with the knowledge and skills to apply them effectively in your own projects.

Many of the design patterns discussed in the Gang of Four's classic *Design Patterns* book relate strictly to object-oriented programming (OOP) in C++. Rust has done an excellent job of making some of those patterns obsolete by providing better alternatives or including them in its standard library (such as iterators). Although the death of OOP has been greatly exaggerated, Rust's abstractions make more intuitive sense when you grok them.

OOP often leads to excessive boilerplate code and overly complex patterns. Sometimes, we justify complexity for the sake of complexity in OOP, engaging in mental gymnastics. Complex systems, however, tend to fail faster and more dramatically than simple systems and are also more challenging to understand.

I find Rust's approach to software design and architecture refreshing, and I hope you do too. Rust's language designers threw away much legacy OOP cruft, focusing instead on what's needed to build quality software. Rust doesn't suffer from the cult of complexity that languages like C++ and Java have fostered.

1.2 *What design patterns are*

Defining *design patterns* is a little tricky—often, a case of knowing it when you see it. The more patterns you learn, the easier it becomes to recognize patterns when you come upon and reimplement them. Learning the most common design patterns will allow you to recognize them immediately and implement them quickly. They are *patterns* because we often see them repeated in many contexts, and they are *design patterns* because they are high-level abstractions that help us design and architect software sensibly.

Some properties of design patterns are common to all patterns and not specific to any particular programming language. These properties are as follows (though this list may not be exhaustive):

- Design patterns are *reusable*.
- Design patterns can be applied widely and broadly.
- Design patterns solve problems in a way that makes it easy to reason about how someone else's code works.
- Design patterns are well understood by other experienced developers.
- Code that doesn't follow well-established patterns may fall under the category of *antipatterns*.

In terms of that last item, you may think, "But hey—I just invented this great new pattern!" Perhaps you did, but until your pattern becomes widely used and understood, it's probably *not* a good idea to expect others to understand or use it. Great design patterns become widely adopted over time and are easy to understand and reason about.

Design patterns should not be adhered to religiously; they provide a familiar template for new software designs while allowing a lot of freedom in terms of implementation details. A good design pattern applies to a wide range of applications while imposing minimal constraints on the author. Design patterns evolve as new language features and paradigms emerge, and the essence of many core patterns has changed little in the past few decades.

In this book, I use broad definitions of *patterns* and *design patterns*. I refer to techniques, idioms, and conventions that are widely used and understood by the Rust community as *patterns*. These patterns can range from big and complex, involving multiple structures and components, to small and simple, consisting of a single function or method. On the other hand, I use the term *design pattern* to encompass widely applicable patterns that serve as templates for code design and solve common programming problems. I use *patterns* and *design patterns* interchangeably throughout this book, but I generally refer to *patterns* as a subset of *design patterns*.

What are antipatterns?

Antipatterns are the evil cousins of design patterns. We usually talk about design patterns as being the right way to solve a certain class of problems; therefore, antipatterns are the wrong way to solve a certain class of problems. This book doesn't discuss antipatterns exhaustively because, for the most part, Rust is designed to make it relatively hard to construct antipatterns in the first place.

Antipatterns are (in most cases) the wrong tool for the wrong job. You wouldn't use a hammer to drive a screw, and you wouldn't use a screwdriver to hammer a nail.

I'll discuss antipatterns in chapter 10. But I'll provide reminders throughout the book to show when you shouldn't use specific patterns.

I should also take a moment to distinguish *patterns* from *idioms* as I define them in this book. A few definitions of the differences have emerged, but I'll focus on two key points: idioms generally relate to the code itself, and patterns generally relate to the design and architecture of your software. Another way to say the same thing: patterns are composed of idioms. Some patterns may also be idioms (they prefer iterators to `for` loops, for example), but an idiom is not a pattern, as using snake case for variable names, for example, is not a pattern. Idioms generally relate to syntax and code formatting, such as naming conventions, code style, and other low-level details.

In a hierarchical sense, we can think of idioms as the lowest level of abstraction, design patterns as the middle level, and the overall architecture as the highest level of abstraction, as shown in figure 1.1. The architecture of any system is composed of many smaller units of design patterns, which are in turn composed of many idioms.

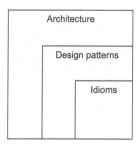

Figure 1.1 Hierarchy of idioms, patterns, and architecture

We can also think about design patterns and programming languages in the same way that we think about spoken and written languages. Languages evolve, new words are created, and old words and phrases go out of style.

If you try to invent your own words or phrases, however, they may seem like nonsense to others. The entire point of languages is to communicate ideas easily, be understood by others, and feel connection to other human beings. In the context of programming, if you decide to reject the software social norms and march to the beat of your own drum, that may be fine, but there's a good chance that other people will struggle to understand your code and won't necessarily want to contribute or work with it. In some cases, that tradeoff is acceptable, but software is often used in social contexts involving customers, users, managers, peers, and so on. No one is an island.

You can't go far in writing about design patterns without mentioning the Gang of Four's *Design Patterns*, well known among programmers as being the original or canonical textbook on design patterns. That book—the full title of which is *Design Patterns: Elements of Reusable Object-Oriented Software*—was written by Erich Gamma, Richard Helm, Ralph Johnson, and John Vlissides and published by Addison-Wesley Professional in 1994; it includes examples written in C++ and Smalltalk.

Some patterns presented in that book have since been added to many programming languages as core features. Perhaps the best examples are iterators, which are part of nearly every programming language and core library because of how useful the iterator pattern is, how well it solves the problem of iterating over elements in a data structure, and how well-understood it is. Implementing iterators from scratch to learn how they work is still fun, but you can use the built-in equivalents in most languages.

Design patterns fit into what I call the three pillars of good software design: algorithms, data structures, and design patterns (figure 1.2). You, as an author of software, need to understand each of these pillars and apply them effectively. Learning design patterns alone is not enough; you also need a good knowledge of algorithms and data structures to build good software.

To summarize, design patterns are high-level abstractions above the core grammar and syntax of a programming language that allow us to communicate ideas effectively and produce high-quality code. Good communication is the responsibility of the person

Figure 1.2   The three pillars of good software design

who delivers the message, not the person who receives it, but it certainly helps if the receiver speaks the same language.

1.3   *Why this book is different*

Since the Gang of Four's *Design Patterns* was published, many more books on design patterns have appeared, and in that sense, this book is no different from those later books. In this book, however, I present some ideas that are specific to Rust. As Rust continues to grow in popularity and proliferation, it's essential to catalog, document, and describe the patterns we use with Rust.

Unlike the Gang of Four's book, this book is not a catalog of design patterns but a discussion and exploration of patterns, examples, and implementations of specific patterns. I don't want to catalog and classify patterns for two reasons: patterns aren't merely templates or boilerplate, and copying and pasting a pattern will get you only about 10% (or less) of the way toward complete code. This book is for readers who have an appetite for knowledge and personal growth.

To use another food analogy, a particular dish (such as lasagna) could be a design pattern; it's part of a considerable dining experience involving multiple courses, drinks, and impeccable service. The real challenge for the chef is deciding how to make their version of a dish, where to source the ingredients, and how to bring everything all together and present the food in an appetizing way. (As anyone who's worked in restaurants knows, presentation is everything.) Programming is both a science and an art; it's a highly creative endeavor that's more than the lines of code. Mimicry gets you only so far.

Rust's unique language features require a little more thought when it comes to API design and the act of building high-quality code. In particular, we have to think harder about how we manage memory and object lifetimes, pass values between contexts, avoid race conditions, and ensure that our APIs are ergonomic. Additionally, Rust is full of greenfield opportunities to create or discover new patterns, which will certainly evolve after this book is published. Before we can go to Mars, we must build a rocket that can take us to Mars and also solve the myriad problems that will arise during the seven-month journey.

Rust is a delightful, wonderful language that is unique partly because of how it evolved—entirely as a community effort. Its abstractions simultaneously unlock new patterns and make old patterns obsolete. Learning the language's syntax is one thing, but to write great Rust code, we need to use the correct patterns in the right places and use them correctly.

1.4 Tools you'll need

This book includes a collection of code samples that are freely available under the MIT license. To obtain a copy of the code, you need an internet-connected computer with a supported operating system (https://mng.bz/JZpa) and the tools shown in table 1.1 installed. For details on installing the tools, see the appendix.

Table 1.1   Required tools

Name	Description
git	The source code for this book is stored in a public repository hosted on GitHub at https://github.com/brndnmtthws/idiomatic-rust-book.
rustup	Rust's tool for managing Rust components. `rustup` will manage your installation of `rustc` and other Rust components.
gcc or clang	You must have a copy of GCC or Clang installed to build certain code samples. Clang is likely the best choice for most people; thus, it's referred to by default. When the `clang` command is specified, you may freely substitute `gcc` if you prefer.

Summary

- Good design patterns are reusable, broadly applicable, and capable of solving common programming problems.
- The hallmarks of a good design pattern are that it becomes widely adopted over time and is easy to understand and reason about.
- An antipattern is a design pattern that's poorly understood, underspecified, or considered harmful.
- This book presents Rust-specific design patterns that take advantage of the unique features provided by the Rust language and its tooling.
- You need an up-to-date installation of Rust, Git, and a modern compiler such as GNU's GCC or LLVM's Clang.
- To get the most out of this book, follow along with the code samples at https://github.com/brndnmtthws/idiomatic-rust-book.

Rust's basic
building blocks

This chapter covers
- Exploring the core Rust patterns
- Diving into Rust generics
- Exploring traits
- Combining generics and traits
- Deriving traits automatically

In this chapter, I'll introduce and discuss some of Rust's most important abstractions and features, which I call *building blocks* and which serve as the foundation of nearly all design patterns in Rust. Reviewing and understanding these building blocks before diving deeper into other patterns is crucial. For some readers, this chapter may appear to be a review of language basics; it sets the stage for more advanced topics, however, so I recommend that you don't skip it.

We'll begin by discussing generics and traits in Rust. They are the core building blocks of nearly every design pattern in Rust, along with Rust's pattern matching and functional features (discussed in chapter 3). These elements constitute the meat and potatoes of the language.

2.1 Generics

After you've moved beyond basic syntax, generics are likely the first big topic you'll need to learn. Rust's generics are compile-time, type-safe abstractions that also enhance metaprogramming; they allow you to use placeholders instead of concrete types in function and structure definitions. Generics (combined with traits, which we'll discuss in section 2.2) permit type-safe programming in a way that doesn't require explicit definitions of every possible type.

Most commonly, we use generics to define structures, functions, and traits that work with any type. You might have a function that works with integers, floats, or strings, and you don't want to write the same function multiple times for each type. Generics let you write the function once and use it with any type.

Generics let you build types that are composed of other types without necessarily needing to know about all possible type combinations or downstream uses. Because generics are compile-time abstractions, you incur no cost or runtime overhead by using them. Generics increase complexity at compile time, however.

Rust's generics are similar to C++'s templates and Java's generics, so if you're coming from those languages, you'll probably feel at home from the start. In C, macros are sometimes used as a way to do generic metaprogramming, but C's macros are *not* type-safe like generics in Rust, C++, and Java.

Some languages bolted on generics as late features, but Rust was (mostly) designed from the start with generics in mind. As a result, generics fit well within the language, are used nearly everywhere, and don't feel kludgy or out of place.

2.1.1 A Turing-complete type system

Rust's type system is Turing-complete, and with generics, you can write programs that execute at compile time, which is a neat trick akin to using the compiler as a CPU. When I say *Turing-complete*, I mean that Rust's type system is capable of expressing any computation that can be computed by a Turing machine—that is, you can compute anything that's computable. Turing completeness in a type system is important because it enables you to compute anything at compile time, as opposed to run time, which unlocks some interesting capabilities.

One example of using types for computation is a Minsky machine implemented with Rust's type system, which you can find at https://github.com/paholg/minsky. A *Minsky machine* is a simple register-based counter machine that is computationally equivalent to a Turing machine, and we can think of a Minsky machine as analogous to a simple CPU. Thus, if we can build a Minsky machine using Rust's type system, we can effectively use Rust's type system to compute anything that's computable.

To get value out of Rust, you don't need to worry much about the Turing completeness of its type system, and in practice, you probably won't need to use the type system for computation. For most people, the main benefits of a Turing-complete type system are the safety and performance features it enables.

2.1.2 *Why generics?*

In statically typed languages like Rust, the compiler needs to know the type of every-thing at compile time. Requiring type information at compile time, before execution, contrasts with dynamically typed languages such as Python and Ruby, which deter-mine the types at run time. Generics allow you to write code that works with any type without the developer's needing to know the type at compile time. Instead, we let the compiler figure out the types.

We employ generics to follow the *DRY* (Don't Repeat Yourself) principle through-out our codebase. Writing the same code in many places where the only difference is the type signature is a recipe for headaches.

The downside to generics is that they can make code harder to read and write, so it's essential to strike a balance between using generics and writing clear, readable code. The difficulty in using generics stems from the fact that we're adding layers of abstraction to our code, particularly abstractions that require additional cognitive load on behalf of the programmer. Also, the compiler can't always figure out the types you want, so you may need to provide hints to tell it what you're trying to do, which may make generics seem troublesome and verbose. Most of the time, how-ever, the additional mental effort required to use generics is worth any perceived short-term suffering, as generics allow you to build more flexible, reusable, and robust software.

2.1.3 *Basics of generics*

Let's explore the syntax of generics. A basic struct with a single generic field looks like this:

```
struct Container<T> {
    value: T,
}
```

Here, we have a basic container that holds a value of type T, which is defined as a generic parameter in angle brackets. Generics can be used in structs, enums, func-tions, impl blocks, and more. You'll encounter this syntax everywhere in Rust. When you see the angle brackets (< ... >), you know you're working with generics.

Creating an instance of a generic struct is relatively easy. Often, the compiler can infer the type parameter automatically:

```
let str_container = Container { value: "Thought is free." };
println!("{}", str_container.value);
```

This container is of type Container<&str>, but we don't need to specify the generic type explicitly because the compiler can infer it.

This code snippet creates a Container<&str> instance called str_container. Run-ning the code prints Thought is free., as expected.

Sometimes, the compiler needs hints to determine the generic type. Suppose we want to store an `Option<String>` in our container but initialize it with `None`. If we try the code

```
let ambiguous_container = Container { value: None };
```

the compiler will fail with the following error:

```
error[E0282]: type annotations needed for `Container<Option<T>>`
 --> src/main.rs:8:50
  |
8 |     let ambiguous_container = Container { value: None };
  |         ------------------       ^^ cannot infer type for type parameter
  |                                     `T` declared on the enum `Option`
  |         |
  |         consider giving `ambiguous_container` the explicit type
  |         `Container<Option<T>>`, where the type parameter `T`
  |         is specified
```

Luckily, the compiler tells us exactly what we need to do. We can update our code like this to let the compiler know that we want to use `Option<String>`:

```
let ambiguous_container: Container<Option<String>> =
    Container { value: None };
```

The only difference is that we're specifying the target type on the left side of the assignment. The types need to match so that the compiler can infer what we're looking for.

Another way to do the same thing is to use the `fn new()` constructor pattern (which we'll revisit in chapter 4), which is often used but not required in Rust:

The generic parameter T appears twice—for the impl block and Container. You can have more complex constructions (such as placeholders, concrete implementations, and default types), but this construction is the simplest generic construction.

We're moving value into the struct. In other words, no references, copies, or cloning are used.

```
impl<T> Container<T> {
    fn new(value: T) -> Self {
        Self { value }
    }
}
```

We can use the short form of assignment here because our local variable value matches the name of value in the struct. The longer equivalent would be value: value.

Then we can call `new()`. This time, however, we tell the compiler what our desired target type is on the right side of the assignment by calling the function explicitly with our target type:

```
let short_alt_ambiguous_container =
    Container::<Option<String>>::new(None);
```

I find this form to be a little cleaner and easier to read in many cases. In some instances, you *must* use this form of assignment because the assignment is still too ambiguous for the compiler to infer the target type. In those cases, the compiler lets you know you need to disambiguate.

As mentioned earlier, generic parameters can be added to all structure and function types in Rust. We can do some neat things with generics, such as constructing recursive structures with generics. As an example, we can create a structure that holds an instance of itself, such as a linked list that includes a generic parameter:

```
#[derive(Clone)]          <──┐  We can implement the Clone
struct ListItem<T>           │  trait automatically by using
where                        │  the #[derive] attribute.
    T: Clone,
{
    data: Box<T>,
    next: Option<Box<ListItem<T>>>,
}
```

We can also use this pattern with enums. Consider this enum, which could be used to construct linked lists (albeit a useless form of them):

```
enum Recursive<T> {
    Next(Box<Recursive<T>>),
    Boxed(Box<T>),
    Optional(Option<T>),
}
```

Here, an enum called `Recursive` can hold a pointer to another `Recursive`, a boxed `T`, or an optional `T`. This example is fairly useless, but it shows what you can do with generics.

> **NOTE** I use the linked-list example throughout the book to demonstrate various Rust features, and I'll build up this example along the way. If you aren't familiar with linked lists, a singly linked list is a data structure consisting of a sequence of elements, each containing a reference to the next element, such as A → B → C → ... → Z.

We could apply this pattern to our linked list by using a structure that looks something like this instead of `Option`:

```
enum NextNode<T> {
    Next(Box<ListNode<T>>),
    End,
}

struct ListNode<T> {
    data: Box<T>,
    next: NextNode<T>,
}
```

Our list node holds a `Box` of `T` and an optional `next` that points to the next node in the list. This code is nice and succinct. For the sake of clarity, however, it's probably better to use `Option` rather than create an equivalent.

> **NOTE** Implementing linked lists in Rust properly is more complicated than I show in this chapter. I'll revisit linked lists later in this book and demonstrate using `Rc` and `RefCell`, which is a better way to construct linked lists. The preceding example wouldn't be useful for most practical applications.

2.1.4 *Exploring Rust's Option*

Let's take a look at Rust's `Option`, the definition of which is as follows:

```
pub enum Option<T> {
    None,
    Some(T),
}
```

Rust's `Option` is one of the most delightful examples of generics in practice. Its definition is simple and elegant, yet it provides an incredibly powerful abstraction.

2.1.5 *Marker structs and phantom types*

Sometimes, you want to make structures with generic parameters, but you don't necessarily want to use the generic parameters in the structure itself. This situation calls for *phantom types*, which enable you to use generic parameters that aren't members of your structure. Phantom types allow the use of patterns such as struct tagging, which we'll discuss in chapter 7.

The following code snippet has a structure that includes a type parameter, but that type is not used within the structure itself (we have only the type information at compile time):

```
struct Dog<Breed> {
    name: String,
}
```

The `Dog` structure holds a dog's name. We want to keep track of the breed of the dog, but we care about those details only at compile time (not run time), so we can effectively store that information as a type parameter and not bother including a `breed:` `Breed` field within the struct. We'll need to create some types to identify our breeds, which we'll do as follows:

```
struct Labrador {}
struct Retriever {}
struct Poodle {}
struct Dachshund {}
```

We're using an empty struct to label each breed. We could use any type, but we'll use an empty struct for this example. Trying to compile the code as is, however, yields the following error:

```
error[E0392]: parameter `Breed` is never used
  --> src/main.rs:27:12
```

```
   |
27 | struct Dog<Breed> {
   |             ^^^^^ unused parameter
   |
   = help: consider removing `Breed`, referring to it in a field,
   or using a marker such as `PhantomData`
   = help: if you intended `Breed` to be a const parameter,
   use `const Breed: usize` instead
```

The compiler is unhappy because we added an unused generic parameter to the struct, which the compiler (rightfully) notes is an error. We can add a phantom field to let the compiler know that we want the parameter, but we only care about the value at compile time and thus don't need to store it in the struct:

```
use std::marker::PhantomData;

struct Dog<Breed> {
    name: String,
    breed: PhantomData<Breed>,
}
```

When we construct a `Dog`, we still need to provide the phantom data, although it will be optimized out at compile time:

```
use std::marker::PhantomData;

let my_poodle: Dog<Poodle> = Dog {
    name: "Jeffrey".into(),
    breed: PhantomData,
};
```

`PhantomData` is a special kind of marker that you'll encounter when working with Rust. Markers are typically used as *marker traits*, but in this case, `PhantomData` is a *marker struct*. The Rust standard library includes several marker traits; we'll discuss marker traits in detail in chapter 7.

One critical use case for marker structs is to specialize distinct types at compile time, which can be useful. We can add specialized implementations of `Dog` for each distinct breed if we choose to do so. We can return the name of the breed without storing that value as state or as a separate field within the structure:

```
impl Dog<Labrador> {
    fn breed_name(&self) -> &str {
        "labrador"
    }
}
impl Dog<Retriever> {
    fn breed_name(&self) -> &str {
        "retriever"
    }
}
```

impl Dog<Labrador> is a concrete specialization for Dog with the Labrador breed type. impl doesn't require the Breed generic parameter because it's a concrete specialization.

We can return the breed name without storing it as a field in the struct. The name will be part of the compiled program's data segment.

```
impl Dog<Poodle> {
    fn breed_name(&self) -> &str {
        "poodle"
    }
}
impl Dog<Dachshund> {
    fn breed_name(&self) -> &str {
        "dachshund"
    }
}
```

For each `impl` block, we're creating a concrete specialization for `Dog` with the given type. We can add as many concrete specializations as we want, and if we're missing one, the compiler will let us know. Note that we don't use `impl<T>` because it's not a generic instantiation; we specialize in a concrete type.

Now we can call `breed_name()` on our `Dog` instance to return the breed name. Note that in the `breed_name()` methods, we don't need to use the `'static` lifetime with our `&str` reference because the methods take `&self`. Thus, the compiler can reasonably conclude that the lifetime of the returned string will match `&self`.

Lifetimes and 'static

Rust's lifetimes are powerful features that allow you to specify how long a reference (or a borrow) is valid. References are equivalent to pointers, but unlike pointers as you may know them from C or C++, you cannot perform arithmetic on references. Lifetimes ensure that references are valid for as long as they are used.

The basic idea behind a lifetime parameter (which begins with the single-quote character) is that it lets you tag a reference with a name that the compiler can use to trace the reference's lifetime through its use. Lifetimes look similar to generic parameters, as they're also specified in angle brackets, but they're not the same. The following structure has a lifetime parameter `<'a>`:

```
struct Dog<'a> {
    name: &'a str,
}
```

In this code, we're specifying that the `name` field of the `Dog` structure contains a reference to a string with a lifetime of `'a`. Specifying the lifetime tells the compiler that the reference must be valid for at least as long as the `Dog` structure is valid.

In Rust, `'static` is a special lifetime that lasts for the duration of the program. All string literals have a `'static` lifetime, so we don't necessarily need to specify a lifetime for them. If you're returning a string literal from a function, you can return it as a `&'static str` if you want to specify the lifetime explicitly.

Including the `'static` lifetime for a string literal is optional, but including it can be advantageous if you're returning a string literal from a function because the lifetime makes it clear that the string literal will be valid for the duration of the program.

Finally, we can test our code as follows:

```
let my_poodle: Dog<Poodle> = Dog {
    name: "Jeffrey".into(),
    breed: PhantomData,
};
println!(
    "My dog is a {}, named {}",
    my_poodle.breed_name(),
    my_poodle.name,
);
```

Running this code produces the following output:

```
My dog is a poodle, named Jeffrey
```

My poodle Jeffrey is correctly identified as a poodle, and we've successfully used a phantom type to specialize our `Dog` structure, so it's unlikely that Jeffrey will have an identity crisis.

2.1.6   *Generic parameter trait bounds*

Before we move on to traits in section 2.2, we have to talk briefly about trait bounds. *Trait bounds* are a feature of generics that allows you to control which types can be used with a particular structure or function by specifying which traits must be implemented. Specifically, trait bounds enable us to specify which features must be available for a given generic type parameter. We can specify multiple trait bounds that apply on a per-parameter basis. Reexamining the linked-list example introduced in section 2.1.3, you'll notice two things about the `ListItem` struct:

- We've derived the `Clone` trait, which allows us to call `clone()` on the struct to copy it.
- We've specified that the generic type `T` must also implement the `Clone` trait, with the `where T: Clone` trait bound.

If we want to require that `Clone` *and* `Debug` be implemented, we use the following code to specify that both traits are required:

```
#[derive(Clone)]
struct ListItem<T>
where
    T: Clone + Debug,
{
    data: Box<T>,
    next: Option<Box<ListItem<T>>>,
}
```

2.2   *Traits*

After spending some time writing Rust and familiarizing yourself with syntax, borrowing, and lifetimes, you soon realize that traits, together with generics, are the bread and butter of Rust programming. Traits are incredibly powerful abstractions that form the foundation of much of Rust's libraries. With that power comes responsibility. Traits come with two significant downsides: trait pollution and trait duplication. We'll discuss how to avoid these problems.

Traits allow you to define shared functionality for Rust types. Instances of types (objects) contain state (such as a struct), and traits define functionality on top of that state in a generic way not tied to any particular type.

Traits aren't unique to Rust. They first appeared in a somewhat obscure programming language called Self. Several other languages offer traits, including Scala, Julia, TypeScript, Kotlin (as interfaces), Haskell (as type classes), and Swift (as protocol extensions).

Although traits are often used to manipulate state, they are distinct from their implementation, which is tied to a particular type. That is, traits themselves are generic, but their implementations are concrete, although they can be derived automatically with the #[derive] attribute. Libraries can export traits, trait implementations, or both.

2.2.1 *Why traits are not object-oriented programming*

Rust is not an object-oriented (OO) programming language, but looking at Rust code, you may think it looks similar in terms of ergonomics. Rust has objects, and objects can have methods. An *object* is an instance of a type, such as a struct or enum, that represents state. Calling methods on an object uses syntax similar to that of OO languages (object.method()). Rust, however, is missing one important feature of OO languages: *inheritance.*

Rust's answer to inheritance is traits. Traits aren't the same as classes (or class inheritance), but they solve a similar set of problems. In object-oriented programming (OOP), you extend objects through inheritance. In trait-based programming, you can add traits on top of any structure or data type, and those traits provide specific features. Object inheritance defines an *is-a* relationship, whereas traits define *functionality.*

To put it another way, when comparing traits with OOP, traits extend or add shared features on top of different kinds of state. Traits are different from classes in that their functionality isn't coupled to particular types (or state). Although it's true that classes in C++ can be made generic with templates, C++'s classes don't make this decoupling easy.

2.2.2 *What's in a trait?*

Traits comprise a definition and any number of optional implementations. A trait definition typically includes these components:

- A trait name
- An optional set of methods (with optional default implementations)
- Optional placeholder generic types
- Optional set of required traits

At a bare minimum, a trait requires only a name, so the following code snippet constitutes a valid trait definition:

```
trait MinimalTrait {}
```

Trait *implementations* apply the definition of the trait to a specific type. We generally write concrete trait implementations for distinct types, but Rust's trait system is flexible enough that we don't have to implement a trait for every possible type. Traits may also use generic data types (discussed in section 2.2.4), which provide another way to specify complex relationships. Although trait implementations are concrete, you can also provide blanket implementations of traits that apply to all types that satisfy the blanket conditions. (We'll discuss blanket implementations in chapter 7.) Following is a basic example of a trait with an implementation in Rust:

```
trait DoesItBark {                    ◁──── The trait definition block
    fn it_barks(&self) -> bool;       ◁──┐
}                                        │ The trait method
                                         │ signature
struct Dog;
                                         ┌ The trait implementation
impl DoesItBark for Dog {             ◁──┤ (or impl) block
    fn it_barks(&self) -> bool {         │
        true                          ◁──┐ We can hardcode returning true
    }                                    │ because dogs do indeed bark.
}
```

Trait definitions can be empty, which allows them to be used for metaprogramming, such as with marker traits. We'll explore advanced use of traits in chapters 7, 8, and 9.

With OOP, features are added through inheritance in a hierarchy (class C <- class B <- class A). With traits, no inheritance structure is imposed; traits can be applied to any type within your crate. Traits may have dependencies specified as trait bounds (i.e., trait B requires that trait A be implemented), but traits with bounds can still be applied to any type that satisfies those bounds.

In OOP, relationships are defined in terms of the objects themselves. In trait programming, relationships are defined in terms of which traits an object implements rather than which object the behavior is implemented for—a subtle but crucial distinction.

> **NOTE** I implore you to avoid thinking about traits in terms of OO concepts such as classes and inheritance, but I have drawn comparisons in this book to help bridge the gap of understanding for those who come from OO backgrounds. Trying to map these concepts 1:1 doesn't make sense in practice; traits require a different approach. It's best to free your mind and discard the gospel of OOP.

2.2.3   Understanding traits by examining object-oriented code

Traits provide a lot more flexibility than inheritance, which requires a bottom-up relationship. (That is, with inheritance, you define shared behavior at lower levels of the hierarchy.) First, consider a sample in C++ that uses an is-a relationship; then we'll examine how to do the same thing in Rust. We'll start by implementing the relationship shown in figure 2.1.

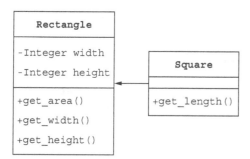

Figure 2.1   Unified Modeling Language
(UML) diagram for C++ geometric shapes

The corresponding C++ code for the UML in figure 2.1 is shown in the following listing.

Listing 2.1   Modeling geometric shapes in C++

```
class Rectangle {          ◁─┐  Models a simple
 protected:                    rectangle with a
  int width;                   width and height
  int height;

 public:
  Rectangle(int width, int height) : width(width), height(height) {}
  int get_area() { return width * height; }
  int get_width() { return width; }
  int get_height() { return height; }
};                                          Models a square, which is merely a rectangle
                                            whose width and height are equal. Thus,
class Square : public Rectangle {    ◁─┘   we can inherit from Rectangle.
 public:
  Square(int length) : Rectangle(length, length) {}
  int get_length() { return width; }
};
```

Writing equivalent code in Rust isn't entirely straightforward; a direct translation to Rust would be awkward. Instead, we'll structure things differently in the Rust version. First, let's examine a listing that models a rectangle.

Listing 2.2   Implementing a rectangle in Rust

```
struct Rectangle {       ◁─┐  Models a simple
    width: i32,              rectangle, which is
    height: i32,            merely width and height
}
```

```
impl Rectangle {
    pub fn new(width: i32, height: i32) -> Self {
        Self { width, height }
    }
}
```

Here, we provide a constructor-like new()
method, which returns a new Rectangle. Creating
new() constructors is a common pattern in Rust.

Next, we'll model a square.

Listing 2.3   Implementing a square in Rust

```
struct Square {
    length: i32,
}
```

Modeling a square is
even simpler; we use
only one attribute.

```
impl Square {
    pub fn new(length: i32) -> Self {
        Self { length }
    }
    pub fn get_length(&self) -> i32 {
        self.length
    }
}
```

Provides a constructor
following the new() pattern

Adds an accessor to fetch the
square's length if we know
that we have a square

Now we can create a `Rectangular` trait.

Listing 2.4   Implementing the `Rectangular` trait

```
pub trait Rectangular {
    fn get_width(&self) -> i32;
    fn get_height(&self) -> i32;
    fn get_area(&self) -> i32;
}
```

Here, we define a Rectangular, which
provides accessors to properties
common to rectangles and squares.

```
impl Rectangular for Rectangle {
    fn get_width(&self) -> i32 {
        self.width
    }
    fn get_height(&self) -> i32 {
        self.height
    }
    fn get_area(&self) -> i32 {
        self.width * self.height
    }
}
```

Implements the
Rectangular trait
for Rectangle

```
impl Rectangular for Square {
    fn get_width(&self) -> i32 {
        self.length
    }
    fn get_height(&self) -> i32 {
        self.length
    }
```

Implements the
Rectangular trait
for Square

```
    fn get_area(&self) -> i32 {
        self.length * self.length
    }
}
```

Figure 2.2 shows the result rendered in UML.

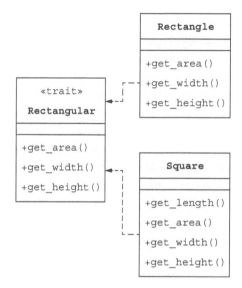

Figure 2.2   UML for Rust
geometric shapes

Last, let's test our code.

Listing 2.5   Testing our `Rectangular` trait

```
fn main() {
    let rect = Rectangle::new(2, 3);
    let square = Square::new(5);

    println!(
        "rect has width {}, height {}, and area {}",
        rect.get_width(),
        rect.get_height(),
        rect.get_area()
    );
    println!(
        "square has length {} and area {}",
        square.get_length(),
        square.get_area()
    );
}
```

The Rust version seems a bit lengthy at first. We have to implement the `Rectangular`
trait twice in a way that appears to violate DRY. But we've done something fundamental:

separated the state (in this case, the dimensions) from the functionality of providing
width, height, area, and so on. As complexity grows, this separation of concerns
scales much better. Running the code in listing 2.5 produces the following output,
as expected:

```
$ cargo run
rect has width 2, height 3, and area 6
square has length 5 and area 25
```

The utility of traits becomes apparent when you consider the complexity of modifying
existing code. In section 2.2.4, we'll explore another example, but we'll approach the
problem in a Rustaceous way.

2.2.4 Combining generics and traits

Suppose that we want to create a function that accepts any type and returns a description of the type. We can write a function that accepts a generic parameter T and
returns a description for that type. We can assume that these types are defined elsewhere, such as Dog and Cat. We must write the descriptions ourselves because the
compiler can't figure them out. To accomplish this task, we'd use a function definition
something like this:

```
fn describe_type<T>(t: &T) -> String { ... }
```

Next, we have to ask ourselves how we get the description of T. The answer is simple:
we need a trait that provides the description. The result looks something like

```
pub trait SelfDescribing {
    fn describe(&self) -> String;
}
```

Great. Now we have a trait that gives us the description of a type. How do we make our
function use that trait? If we try this code, it won't work:

```
fn describe_type<T>(t: &T) -> String {
    t.describe()
}
```

The compiler gives us the following error:

```
error[E0599]: no method named `describe` found for reference `&T` in
the current scope
 --> src/main.rs:6:7
  |
6 |     t.describe()
  |       ^^^^^^^^ method not found in `&T`
  |
  = help: items from traits can only be used if the type parameter is
  bounded by the trait
help: the following trait defines an item `describe`, perhaps you need to
```

```
restrict type parameter `T` with it:
  |
5 |  fn describe_type<T: SelfDescribing>(t: &T) -> String {
  |                   ~~~~~~~~~~~~~~~~~~

For more information about this error, try `rustc --explain E0599`.
```

That's neat! The compiler tells us exactly what to do. We need to instruct the compiler that we want to use the describe() method from the SelfDescribing trait, which we do by creating a trait bound. Trait bounds let the compiler know that a given type must provide an implementation of a particular trait. You'll see trait bounds frequently in Rust; they're often used with generics.

Note that we have two ways to specify the trait bound: inline (as in the compiler error output) or in the explicit where clause, which follows the function definition. Here's what the inline form looks like:

```
fn describe_type<T: SelfDescribing>(t: &T) -> String {
    t.describe()
}
```

Although the inline form is short and sweet, I prefer the where form when the bounds are complex, as it's a bit easier to read as a developer:

```
fn describe_type<T>(t: &T) -> String
where
    T: SelfDescribing,
{
    t.describe()
}
```

Now our code compiles. Let's create some code to test it:

```
struct Dog;
struct Cat;

fn main() {
    let dog = Dog;
    let cat = Cat;
    println!("I am a {}", describe_type(&dog));
    println!("I am a {}", describe_type(&cat));
}
```

Trying to compile this code produces an error because it's missing an implementation:

```
error[E0277]: the trait bound `Dog: SelfDescribing` is not satisfied
  --> src/main.rs:15:41
   |
15 |     println!("I am a {}", describe_type(&dog));
   |                          ------------- ^^^^ the trait `SelfDescribing`
```

```
    is not implemented for `Dog`
  |                         |
  |                         required by a bound introduced by this call
  |
note: required by a bound in `describe_type`
  --> src/main.rs:5:21
  |
5 | fn describe_type<T: SelfDescribing>(t: &T) -> String {
  |                     ^^^^^^^^^^^^^^ required by this bound in
   `describe_type`

error[E0277]: the trait bound `Cat: SelfDescribing` is not satisfied
  --> src/main.rs:16:41
   |
16 |     println!("I am a {}", describe_type(&cat));
   |                          ------------- ^^^^ the trait `SelfDescribing`
    is not implemented for `Cat`
   |                                    |
   |                                    required by a bound introduced by this call
   |
note: required by a bound in `describe_type`
  --> src/main.rs:5:21
   |
5 | fn describe_type<T: SelfDescribing>(t: &T) -> String {
  |                     ^^^^^^^^^^^^^^ required by this bound in
   `describe_type`

For more information about this error, try `rustc --explain E0277`.
```

Again, the compiler tells us precisely what we're missing. (We have to implement SelfDescribing for Dog and Cat.) Let's add the implementations:

```
impl SelfDescribing for Dog {
    fn describe(&self) -> String {
        "happy little dog".into()
    }
}

impl SelfDescribing for Cat {
    fn describe(&self) -> String {
        "curious cat".into()
    }
}
```

Now running our code prints the following:

```
$ cargo run
I am a happy little dog
I am a curious cat
```

One thing to note about the code is that it requires an instance of a type in our trait with the &self parameter on fn describe(&self). Can we do this without requiring an instance of a type? Let's try. We'll modify our trait like so:

```
pub trait SelfDescribing {
    fn describe() -> String;
}
```

Here, we've dropped `&self` from the `describe()` method. Now we'll have to update our `describe_type()` function

```
fn describe_type<T: SelfDescribing>() -> String {
    T::describe()
}
```

and the implementations (by dropping the `&self` parameter):

```
impl SelfDescribing for Dog {
    fn describe() -> String {
        "happy little dog".into()
    }
}

impl SelfDescribing for Cat {
    fn describe() -> String {
        "curious cat".into()
    }
}
```

Last, we change the call to `describe_type()`:

```
fn main() {
    println!("I am a {}", describe_type::<Dog>());
    println!("I am a {}", describe_type::<Cat>());
}
```

Both forms are valid but serve different use cases. If we require `&self` in the method call, we must have an instance of a type to describe it, whereas if we omit the `&self` parameter, we can describe a type without having an object instance.

When you have a basic handle on traits, you can start to apply them to a variety of problems. The most common use of traits is to allow *generic functionality*—shared behavior across types. This use case, however, is the tip of the iceberg, as you can build on traits to create fairly elaborate compile-time patterns, discussed in chapters 7, 8, and 9.

Traits are fun, but they need to be used appropriately. My two biggest problems with traits are trait pollution and trait duplication. *Trait pollution* occurs when you have too many traits. *Trait duplication* occurs when multiple traits provide the same (or similar) functionality. Common programming patterns probably have an existing trait, and whenever possible, it's best to reuse or build atop existing traits. Third-party libraries often define their own traits and sometimes even competing traits, and you can spend a lot of time writing glue code to bridge your code, one library's traits, and another library's traits.

2.2.5 *Deriving traits automatically*

If you're new to Rust, you should familiarize yourself with the commonly used traits in the standard library, including `Clone`, `Debug`, `Default`, iterator traits, and equality traits. Rust also has special traits such as `Drop`, which provides a destructor, and traits that the compiler derives automatically, such as `Send` and `Sync`. You can find a full list of special traits in the Rust language reference at https://mng.bz/wxKa.

For some of the most common traits, you'll use the `#[derive]` attribute to provide implementations automatically. It's common to see struct definitions that use `#[derive]` to derive traits and boilerplate automatically. The following example shows `Clone`, `Debug`, and `Default` with our `Pumpkin` struct:

```
use std::fmt::Debug;

#[derive(Clone, Debug, Default)]
struct Pumpkin {
    mass: f64,
    diameter: f64,
}
```

In this example, we have a `Pumpkin` that can be formatted as a string with `Debug` and cloned with `Clone` and can create a default zeroed instance with `Default`:

```
fn main() {
    let big_pumpkin = Pumpkin {
        mass: 50.,
        diameter: 75.,
    };
    println!("Big pumpkin: {:?}", big_pumpkin);
    println!("Cloned big pumpkin: {:?}", big_pumpkin.clone());
    println!("Default pumpkin: {:?}", Pumpkin::default());
}
```

Running this code prints the following:

```
$ cargo run
Big pumpkin: Pumpkin { mass: 50.0, diameter: 75.0 }
Cloned big pumpkin: Pumpkin { mass: 50.0, diameter: 75.0 }
Default pumpkin: Pumpkin { mass: 0.0, diameter: 0.0 }
```

In practice, you'll need to provide these traits often, as they're widely used throughout the Rust standard library and third-party libraries. Fortunately, this task is easy with `#[derive]`. In the `Option` definition in the Rust standard library, we see the following:

```
#[derive(Copy, PartialEq, PartialOrd, Eq, Ord, Debug, Hash)]
pub enum Option<T> {
    None,
    Some(T),
}
```

Option provides trait implementations for Copy, PartialEq, PartialOrd, Eq, Ord, Debug, and Hash. You may notice that Clone is missing; it's implemented without #[derive].

You don't have to derive your trait implementations, which happens to be the easiest way much of the time; you can always write your own implementations. Suppose you want your default Pumpkin to have a diameter of 5 and mass of 2. You would drop the Default from #[derive] and add the following implementation:

```
impl Default for Pumpkin {
    fn default() -> Self {
        Self {
            mass: 2.0,
            diameter: 5.0,
        }
    }
}
```

Rerunning the code produces the following:

```
$ cargo run
Big pumpkin: Pumpkin { mass: 50.0, diameter: 75.0 }
Cloned big pumpkin: Pumpkin { mass: 50.0, diameter: 75.0 }
Default pumpkin: Pumpkin { mass: 2.0, diameter: 5.0 }
```

2.2.6 *Trait objects*

Rust has a neat feature called *trait objects*, which lets us manage objects as traits instead of as types. You can think of trait objects as behaving similarly to virtual methods in C++ or Java, but they're not the same as inheritance. In terms of implementation details, Rust uses a *vtable* to implement trait objects under the hood, which is a lookup table generated by the compiler to enable dynamic dispatch at run time.

Some people in the Rust community consider trait objects, dynamic dispatch, and vtables to be a form of run-time polymorphism. In some cases, using dynamic dispatch could be viewed as an antipattern, which we'll discuss in chapter 10. I view trait objects as tools, and like all tools, they can be used for good or evil at the behest of the programmer.

We can identify trait objects by using the dyn keyword, and rather than using a type name, we supply a trait. Suppose that we want to store any type within a container. We can do so as long as all the types implement some trait that you specify, as in this example:

```
trait MyTrait {
    fn trait_hello(&self);
}

struct MyStruct1;

impl MyStruct1 {
    fn struct_hello(&self) {
```

```
        println!("Hello, world! from MyStruct1");
    }
}

struct MyStruct2;

impl MyStruct2 {
    fn struct_hello(&self) {
        println!("Hello, world! from MyStruct2");
    }
}

impl MyTrait for MyStruct1 {
    fn trait_hello(&self) {
        self.struct_hello();
    }
}

impl MyTrait for MyStruct2 {
    fn trait_hello(&self) {
        self.struct_hello();
    }
}
```

In this code, we declare `MyTrait`, which provides the `trait_hello()` method. That method is implemented for both `MyStruct1` and `MyStruct2`, which in turn call their own separate `struct_hello()` methods, which print `Hello, world!` Now we can test the code as follows:

```
let mut v = Vec::<Box<dyn MyTrait>>::new();          Adds an instance of
                                                     MyStruct1 to our vector

v.push(Box::new(MyStruct1 {}));          ◀————————   Adds an instance of
v.push(Box::new(MyStruct2 {}));          ◀————————   MyStruct2 to our vector

v.iter().for_each(|i| i.trait_hello());          ◀——   Calls the trait_hello()
// v.iter().for_each(|i| i.struct_hello()); error!  ◀—   method for each trait
                                                         object element in our
            Trying to call the struct_hello() method     vector
            from our structs doesn't work.
```

Running the test code produces the following output:

```
Hello, world! from MyStruct1
Hello, world! from MyStruct2
```

We can't store a trait as an object directly because trait objects are unsized (they don't implement the `Sized` trait). In other words, we need to store our objects in some container type that can hold objects that don't implement `Sized`. That list includes the smart pointers `Box`, `Rc`, `Arc`, `RefCell`, and `Mutex`. We cannot, however, store an unsized object directly within a `Vec`. `Box` (and the other smart pointers) have `where T: ?Sized` in their trait bounds, which means that `Sized` is optional (thus, it can hold trait objects). In Rust, by default, for any generic type `T`, the `Sized` trait is required (equivalent to `where T: Sized`).

   We could not have `Vec<dyn MyTrait>`, for example, because `Vec` does not know how to create unsized objects. A `Box`, on the other hand, decouples allocation from the containment of the element. That is, when we create an object with `Box`, we provide the concrete type at the time of construction; then the compiler can automatically cast the object to the trait object type (that is, from `Box<MyStruct1>` to `Box<dyn MyTrait>`) when we pass or assign the object.

TIP   For more details on trait objects, see the Rust language reference at https://mng.bz/qOp6.

Downcasting trait objects

Aside from the overhead of vtables, one limitation of trait objects is that we can call methods only on the trait, not the concrete type. If we want to coerce a trait object into a concrete type, we can do so by using a downcast. We can use `Box`, `Rc`, and `Arc` to perform a downcast, and the `Any` trait provides a method to downcast. If we want to obtain a reference, however, we need to use `Any`; the `downcast()` method on `Box`, `Rc`, and `Arc` will consume the object, but `Any` provides `downcast_ref()`, which returns a reference.

The `Any` trait is derived automatically for any types that have a `'static` bound, which means that they are free of nonstatic references, so this trick works only for objects that are `dyn Any + 'static`.

To get an `Any` object on our trait object, we must first provide a way to get the `Any` object out from inside the `Box`. We can't simply call `downcast_ref()` on `Box<dyn MyTrait>` because `Box` itself implements `Any`, and we'll get the wrong object. Instead, we have to add an `as_any()` method to our trait to give us the inner object. We can update our code like so:

```
trait MyTrait {
    fn trait_hello(&self);
    fn as_any(&self) -> &dyn Any;        ⟵┐ This trait method
}                                              provides a way to
                                               get &dyn Any.

impl MyTrait for MyStruct1 {
    fn trait_hello(&self) {
        self.struct_hello();
    }
    fn as_any(&self) -> &dyn Any {
        self                             ⟵┐ Returns an instance
    }                                        of Any for self
}

impl MyTrait for MyStruct2 {
    fn trait_hello(&self) {
        self.struct_hello();
    }
```

(continued)

```
    fn as_any(&self) -> &dyn Any {
        self                          Returns an instance
    }                                 of Any for self
}
```

Now we can obtain a reference to the original object type:

> We could also use into_iter() here rather than iter(). In the full code sample, this is the last time we use the v object; thus, we can consume it rather than use a reference via iter().

```
println!("With a downcast:");
v.iter().for_each(|i| {
    if let Some(obj) = i.as_any().downcast_ref::<MyStruct1>() {
        obj.struct_hello();
    }
    if let Some(obj) = i.as_any().downcast_ref::<MyStruct2>() {
        obj.struct_hello();
    }
});
```

Last, several crates provide more advanced downcasting features, such as `downcast`, `downcast-rs`, and `Anyhow`. I discuss crates in detail in chapter 4.

One final note on dynamic dispatch: you should think carefully about whether you want to use traits this way. You probably shouldn't abuse this feature to implement OO-style polymorphism, for example; we discuss it as an antipattern in chapter 10.

No definitive guide to Rust's core traits exists, but an excellent place to start is the prelude documentation at https://doc.rust-lang.org/std/prelude/index.html, which lists the traits and types available in the default Rust namespace.

Last, you can't implement external traits for types outside your crate, but you can work around this situation with wrapper structs or extension traits, which we'll explore in chapters 5 and 7. You can still implement local traits (traits defined within your crate) for any type, even those from external crates. You can implement external traits with multiple type parameters for external types so long as one of the covered type parameters is a local type. For details, see the Rust language reference on orphan rules at https://mng.bz/7dA7.

Summary

- Generics are key abstractions in Rust that enable type-safe code reuse.
- Generics let us include type parameters when defining structs, enums, and functions to create objects and functions that can handle many types of values rather than one specific concrete type.
- Commonly, generics are used to create container types (those that contain other kinds of arbitrary data).
- Traits allow us to add shared functionality on top of different types in Rust.

- We can combine generics and traits to build small libraries that perform their functions well rather than large applications or libraries.
- When we define generic parameters, we can specify which traits they must implement with trait bounds so we can build generic code that depends on shared behavior without specifying concrete types.
- We can derive traits automatically by using `#[derive(…)]`, which saves a lot of typing and boilerplate.

Code flow

This chapter covers

- Discussing pattern matching
- Handling errors with pattern matching
- Reviewing Rust's functional programming patterns

We need to continue to review more of Rust's core language features—its building blocks—before diving into design patterns. In this chapter, we'll start by discussing pattern matching and functional programming. *Pattern matching* allows us to control the code flow, unwrap or destructure values, and handle optional cases. *Functional programming* lets us build software around the unit of a function, which is one of the most basic and easiest-to-understand abstractions.

These building blocks are distinct but can be combined in many ways to create new abstractions. We'll tie these building blocks together to create more elaborate design patterns by combining them in various ways. In cooking (to use an analogy), we employ four essential elements in different combinations from multiple sources to create delicious foods: salt, fat, acid, and heat. Before making patterns based on these elements, we must understand them in depth.

3.1 A tour of pattern matching

Up to now, we've discussed generics and traits that make up Rust's core compile-time features. *Pattern matching* is a run-time feature that enables a variety of lovely code flow patterns. We can match types, values, enum variants, and more. Rust's pattern matching is powerful because it supports several kinds of matching (on both values and types); most important, it enables clean, functional programming patterns.

> **NOTE** *Pattern matching* is not to be confused with *design patterns*. Pattern matching is a core language feature of Rust (and other languages), and although we can use it to build design patterns, it isn't strictly a design pattern.

If you've used a switch/case statement, Rust's pattern matching will look familiar. But Rust's pattern matching is much more potent than a switch/case statement. Some languages provide an equivalent feature, but pattern matching is still somewhat niche, and many mainstream languages do not have it. Pattern matching likely saw its first widespread use in Prolog and is an essential feature of functional languages such as Haskell, Scala, Erlang (itself influenced by and initially implemented in Prolog), Elixir, and OCaml.

A basic pattern match starts with the `match` keyword, which makes it easy to recognize. As with a switch/case statement, we list all the patterns we want to match with an optional catch-all at the end. In Rust, however, we have to match all possible patterns or provide the catch-all case. The Rust compiler tells us if we're missing a case with an error.

3.1.1 Basics of pattern matching

A simple example of pattern matching is unwrapping an `Option` and printing whether it contains a value.

Listing 3.1 Pattern matching an `Option`

```
fn some_or_none<T>(option: &Option<T>) {
    match option {
        Some(_v) => println!("is some!"),
        None => println!("is none :("),
    }
}
```

We unwrap the option's value into _v. Prefixing a variable with an underscore tells the compiler that the value is unneeded.

Unwrapping `Option`, `Result`, or other structures that contain optional data is a common use of pattern matching. Using pattern matching to unwrap data is arguably the killer feature of pattern matching because the compiler requires us to handle all cases. It takes the guesswork out of knowing whether you've handled all possible cases. Pattern matching cannot guarantee that your code is free of logic errors; instead, it makes code easier to reason about.

An astute reader may notice that in listing 3.1, we discarded the value of `Some(_v)`, but it would be nice to print its value instead. To do so, we need to use a binding in

our pattern match and update the generic parameter `T` to include a trait bound for
`std::fmt::Display`.

```
fn some_or_none_display<T: std::fmt::Display>(option: &Option<T>) {
    match option {
        Some(v) => println!("is some! where v={v}"),
        None => println!("is none :("),
    }
}
```

Now we can call `some_or_none_display()` with an `Option` that contains any value that
implements `std::fmt::Display` and print the value if it's `Some`.

Sourcing security vulnerabilities

The vast majority of critical security vulnerabilities in software tend to involve the
same class of problems: memory safety. An analysis by Microsoft (http://mng.bz/
yZKy) found that 70% of security vulnerabilities in Microsoft products involved memory safety bugs in C and C++ code. Examples of memory safety problems include

- Reading/writing outside the bounds of an array
- Dereferencing invalid pointers, such as null pointers
- Using memory after it's been freed
- Attempting to free memory that was previously freed (such as double-free)
- Failing to handle error cases

Rust's safety features seek to eliminate these cases, and pattern matching is a key
feature that helps programmers avoid common pitfalls by requiring that *all* cases be
handled. Pattern matching on an `Option` into `Some` and `None` is a good example of
how Rust forces us to handle all possible cases.

Choosing Rust for critical software is akin to buying an insurance policy or a put con-
tract (a financial instrument that protects against catastrophic loss). Rust is a way to
hedge against the risk of security vulnerabilities and protect your users and your reputation. The premiums you pay are the time and effort involved in learning Rust's
safety features, the discipline required to use them, and any additional cognitive load
on your part. The payout is peace of mind from knowing that your software is less
likely to be the next headline in a security breach. The simple tradeoff is a little extra
work upfront for a lot less work later should things go wrong.

Pattern matching isn't limited to unwrapping `Option` types, although that use case is
common. We can also match specific integral values, including ranges:

```
fn what_type_of_integer_is_this(value: i32) {
    match value {
        1 => println!("The number one number"),
        2 | 3 => println!("This is a two or a three"),
```

```
            4..=10 => println!("This is a number between 4 and 10 (inclusive)"),
            _ => println!("Some other kind of number"),
        }
}
```

Pattern matching is often used to destructure structs, tuples, and enums. You can destructure tuples partially or pull out each element, which can be a convenient way to access inner elements in some cases:

```
fn destructure_tuple(tuple: &(i32, i32, i32)) {
    match tuple {
        (first, ..) => {                                    Matches only on the
            println!("First tuple element is {first}")      first element in a
        }                                                    tuple of any length
    }
    match tuple {
        (.., last) => {                                     Matches only on the
            println!("Last tuple element is {last}")        last element in a tuple
        }                                                    of any length
    }
    match tuple {
        (_, middle, _) => {                                 Matches the middle
            println!(                                        element on a tuple
                "The middle tuple element is {middle}"       with three elements
            )
        }
    }                                                        Matches every
    match tuple {                                            element of a tuple
        (first, middle, last) => {                           with three elements
            println!("The whole tuple is ({first}, {middle}, {last})")
        }
    }
}
```

You can have multiple equivalent `match` expressions, but the block always returns the expression from the first matching pattern. In the preceding example, we use a separate `match` block for each case because all matches are valid. If you have multiple equivalent patterns in a `match` block, your code will compile but produce a warning, like the following code snippet:

```
fn unreachable_pattern_match(value: i32) {
    match value {
        1 => println!("This value is equal to 1"),
        1 => println!("This value is equal to 1"),
        _ => println!("This value is not equal to 1"),
    }
}
```

Compiling this code will produce the following warning for the second `match` case:

```
warning: unreachable pattern
  --> src/main.rs:56:9
```

```
56 |            1 => println!("Second match: This value is equal to 1"),
   |            ^
   |
   = note: `#[warn(unreachable_patterns)]` on by default
```

A guard allows you to match conditionally by using an `if` statement after the pattern, which can use the matched value or a separate value passed to the guard. The following code uses a guard to match on a value and a Boolean:

```
fn match_with_guard(value: i32, choose_first: bool) {
    match value {
        v if v == 1 && choose_first => {
            println!("First match: This value is equal to 1")
        }
        v if v == 1 && !choose_first => {
            println!("Second match: This value is equal to 1")
        }
        v if choose_first => {
            println!("First match: This value is equal to {v}")
        }
        v if !choose_first => {
            println!("Second match: This value is equal to {v}")
        }
        _ => println!("Fell through to the default case"),
    }
}
```

You can't match values of different types within a `match` statement. All match cases or branches within the same `match {}` block should apply to the same type. The `match` block is an expression, so each branch (and each expression therein) needs to return the same type. You can unwrap structures that contain different types (such as an enum), but you can't match generically. The following code, for example, is not valid:

```
fn invalid_matching<T>(value: &T) {
    match value {
        "is a string" => println!("This is a string"),
        1 => println!("This is an integral value"),
    }
}
```

Attempting to compile this code will produce the following compiler output:

```
error[E0308]: mismatched types
  --> src/lib.rs:3:9
   |
1  | fn invalid_matching<T>(value: &T) {
   |                     - this type parameter
2  |     match value {
   |           ----- this expression has type `&T`
3  |         "is a string" => println!("This is a string"),
   |         ^^^^^^^^^^^^^ expected `&T`, found `&str`
```

```
    |
    = note: expected reference `&T`
               found reference `&'static str`

error[E0308]: mismatched types
 --> src/lib.rs:4:9
    |
1 |  fn invalid_matching<T>(value: &T) {
    |                          - this type parameter
2 |      match value {
    |            ----- this expression has type `&T`
3 |          "is a string" => println!("This is a string"),
4 |          1 => println!("This is an integral value"),
    |          ^ expected type parameter `T`, found integer
    |
    = note: expected type parameter `T`
                    found type `{integer}`

For more information about this error, try `rustc --explain E0308`.
```

We can destructure different inner types if we use an enum. `DistinctTypes` allows us to match distinct named types in `match_enum_types()`, just as you would an `Option`:

```
enum DistinctTypes {
    Name(String),
    Count(i32),
}

fn match_enum_types(enum_types: &DistinctTypes) {
    match enum_types {
        DistinctTypes::Name(name) => println!("name={name}"),
        DistinctTypes::Count(count) => println!("count={count}"),
    }
}
```

We can destructure structs to extract specific values and even match on particular values within a struct, as I'll demonstrate in the following example. This code snippet creates an enum for cat colors, a struct that contains the cat's name and its color, and a function `match_on_black_cats()` that prints the cat's name and tells us whether it's a black cat:

```
enum CatColor {
    Black,
    Red,
    Chocolate,
    Cinnamon,
    Blue,
    Cream,
    Cheshire,
}

struct Cat {
    name: String,
```

```
        color: CatColor,
    }

fn match_on_black_cats(cat: &Cat) {
    match cat {
        Cat {
            name,
            color: CatColor::Black,
        } => println!("This is a black cat named {name}"),
        Cat { name, color: _ } => println!("{name} is not a black cat"),
    }
}
```

We can quickly test the code as follows:

```
let black_cat = Cat {
    name: String::from("Henry"),
    color: CatColor::Black,
};
let cheshire_cat = Cat {
    name: String::from("Penelope"),
    color: CatColor::Cheshire,
};
match_on_black_cats(&black_cat);
match_on_black_cats(&cheshire_cat);
```

Running the preceding test prints the following output:

```
This is a black cat named Henry
Penelope is not a black cat
```

3.1.2   *Clean matches with the ? operator*

Pattern matching is an excellent way to handle errors, but code can get messy when we have too many matches or matches that are too deeply nested. We can combine pattern matching with the ? operator to handle functions that return `Result` or `Option` cleanly by returning immediately when `Result` or `Option` returns an error or `None`, respectively. To use the ? operator, we need to be inside a function that returns `Result` or `Option`. The ? operator allows us to flatten our code considerably, which improves readability:

> Our function returns a std::io::Result, which is a type alias for Result with the std::io::Error error type provided for convenience. The return payload is a unit ().

```
fn write_to_file() -> std::io::Result<()> {
    use std::fs::File;
    use std::io::prelude::*;

    let mut file = File::create("filename")?;
    file.write_all(b"File contents")?;
    Ok(())
}
```

> All calls to functions returning a Result use the ? operator to denote that in case of an error, the function should return that error.

> We return the unit type with Ok to show success.

```
fn try_to_write_to_file() {
    match write_to_file() {                          Calls our function and
        Ok(()) => println!("Write succeeded"),       matches on the result
        Err(err) => println!("Write failed: {}", err.to_string()),
    }
}
```

In the preceding code, we wrap the call to `write_to_file()` within a pattern-matching expression. If the function returns `Ok(())`, we print `Write succeeded`. In the case of an error, we print `Write failed: …` with the error message.

Using the `?` operator is a super-handy way to keep your code clean by using `Result`. Notice that I used the unit type `()`, a special type in Rust that is essentially a placeholder that carries no value and is optimized out by the compiler. The unit type `()` is often referred to simply as *unit*. The equivalent code without `?` looks something like this example, which includes duplicate code for printing the error case:

```
fn write_to_file_without_result() {
    use std::fs::File;
    use std::io::prelude::*;

    let create_result = File::create("filename");
    match create_result {
        Ok(mut file) => match file.write_all(b"File contents") {
            Err(err) => {
                println!("There was an error writing: {}", err)
            }
            _ => println!("Write succeeded"),
        },
        Err(err) => println!(
            "There was an error opening the file: {}",
            err
        ),
    }
}
```

If we want to chain lots of calls by using the `?` operator, we need to pay attention to their return types. The `?` operator works only with functions that return either a `Result<T, E>` or `Option<T>` that matches the type of the statement with the `?` applied. For `Result<T, E>`, the error types of all the functions using `?` must match the parent function or provide an implementation of the `From` trait so that they can be converted to the target error type. For this reason, you'll often have to write `impl From for … {}` for conversion between error types.

> **TIP** When you're chaining the `?` operator, you can use a few handy methods for converting between `Result` and `Option`, in addition to implementing the `From` trait. For `Result<T, E>`, you can use the `ok()` method to map to `Option<T>`, `err()` to map to `Option<E>`, and `map_err()` to map an error to a different type. For `Option<T>`, use `ok_or()` to map to `Result<T,E>`.

In the preceding example, if we want to use our own error type instead of std::io::Error, perhaps because we want to add more information to the original error, we need to do something like this:

```
enum ErrorTypes {
    IoError(std::io::Error),
    FormatError(std::fmt::Error),
}

struct ErrorWrapper {
    source: ErrorTypes,
    message: String,
}
```

Next, we need to implement From<std::io::Error> for our error wrapper:

```
impl From<std::io::Error> for ErrorWrapper {
    fn from(source: std::io::Error) -> Self {
        Self {
            source: ErrorTypes::IoError(source),
            message: "there was an IO error!".into(),
        }
    }
}
```

Now we can update our file-writing code to use our error type by returning Error-Wrapper in our write_to_file() function:

```
fn write_to_file() -> Result<(), ErrorWrapper> {        Returns a plain Result
    use std::fs::File;                                   instead of std::io::Result
    use std::io::prelude::*;                             using our error type

    let mut file = File::create("filename")?;
    file.write_all(b"File contents")?;
    Ok(())
}

fn try_to_write_to_file() {
    match write_to_file() {                              Prints our error
        Ok(()) => println!("Write succeeded"),           message instead of
        Err(err) => {                                    the one provided
            println!("Write failed: {}", err.message)    by std::io::Error
        }
    }
}
```

If we call our try_to_write_to_file() function, it should (under normal circumstances) print Write succeeded. But in the case of an error (such as not having permission to write a file), the function will print Write failed: ... with the error message provided by File.

Handling errors this way is fairly common in Rust and can save a great deal of typing. This approach is a relatively simple way to integrate errors from third-party crates into your error-handling code. Chapter 4 revisits the `?` operator and error handling in Rust.

3.2 *Functional Rust*

So far, this book has covered the basics: generics, traits, and pattern matching. Now we'll move on to Rust's functional features, including one of my favorite subjects: functional programming. The two core features of functional programming in Rust are *closures* and *iterators*.

Many people have probably used closures and iterators at some point, as they've become trendy. The JavaScript and TypeScript languages and their libraries, for example, make heavy use of closures. Iterators are so common that most people don't think of them as abstractions but as a core feature of all modern programming languages.

Functional programming is a paradigm wherein programs are composed of declarative functions, and mutation of state is discouraged (though not necessarily disallowed, depending on the strictness of the language). Some languages are strictly functional, which means that you're not allowed to change state; the only way to affect state is to use a function that maps one value to another. Also, functional languages discourage side effects, which are actions within a function that might have nondeterministic results, such as I/O or mutating local state.

To support functional programming, some languages have features explicitly designed around functions and handling immutable state. Although Rust is not strictly functional, it encourages functional patterns by making mutability opt-in (with the `mut` keyword) rather than opt-out and by providing core functional features such as closures and iterators.

Functional programming is a wide subject, so I'll stick to reviewing the high-level features in Rust. For a deep dive into functional programming, *Grokking Functional Programming* by Michał Płachta (https://www.manning.com/books/grokking-functional -programming) provides an excellent overview.

3.2.1 *Basics of functional programming in Rust*

Let's jump in by looking at a simple (but not pure) closure:

```
let bark = || println!("Bark!");
bark();
```
Calling println!() introduces side effects because it's an I/O operation, meaning this closure is not pure.

Here, we have a function that barks like a dog with `"Bark!"` It doesn't look like a function because it has no arguments, and the braces have been removed, as they're not necessary. In Rust, closures begin with a list of arguments between two pipes, `||`, followed by a code block. In the case of a single-line function, you can omit the braces (`{}`) for the block. Let's add a parameter to make the function look more function-like:

```
let increment = |value| value + 1;
increment(1);
```

Here, the function takes an integer `value` and returns that value plus 1. We don't need to specify the type of the `value` parameter because the compiler can infer it. Let's make a closure that looks even more function-like by using a code block:

```
let print_and_increment = |value| {
    println!("{value} will be incremented and returned");
    value + 1
};
print_and_increment(5);
```

These examples aren't too interesting. Closures start to get interesting when we talk about *higher-order functions*, which take other functions as parameters. In Rust, you may have encountered higher-order functions when working with iterators, specifically when using `map()`, `for_each()`, `find()`, `fold()`, and similar methods. Higher-order functions are a convenient way to delegate operations to the caller of the function by allowing the caller to supply inner logic to the callee. Closures make the syntax more convenient, delightful, and flexible. The following simple example of using a higher-order function creates an adder that gets its values from other functions:

```
let left_value = || 1;                          A closure that returns 1 and
let right_value = || 2;                         provides impl Fn() -> i32
let adder = |left: fn() -> i32,
             right: fn() -> i32| {              A closure that returns 2 and
    left() + right()                            provides impl Fn() -> i32
};
println!(
    "{} + {} = {}",                             A closure that takes two functions
    left_value(),                               and adds their results together,
    right_value(),                              providing impl Fn(fn() -> i32,
    adder(left_value, right_value)              fn() -> i32) -> i32
);
```

The preceding example has two closures, assigned to `left_value` and `right_value`, respectively, that return a hardcoded integer. Then we create this `adder`, which takes two parameters of type `fn()` → `i32`, a special function type. We can pass any function that matches the signature to the adder. In this case, we add the left and right values together, which is 1 + 2, so our function returns 3. Running this code produces the following output:

```
1 + 2 = 3
```

You can experiment by changing the values returned by `left_value` and `right_value`; you'll see the output change accordingly. You can also try changing the adder to multiply the values instead of adding them.

3.2.2 *Closure variable capture*

If we want to call our adder with a function that doesn't have the proper signature, we could wrap it with another closure to get the correct signature. Let's discuss variable capture in closures to understand why we might need to do this.

Rust provides three traits that aid in functional programming: `Fn`, `FnMut`, and `FnOnce`. These traits are implemented automatically when possible and summarized as follows:

- `Fn` is for functions in the form of `Fn(&self)`, which can be called repeatedly, as they don't consume the variables they capture. All arguments are immutable.
- `FnMut` is for mutable functions, such as those of the form `FnMut(&mut self)`. They can be called repeatedly, as they don't consume the variables they capture, but they do contain mutable references.
- `FnOnce` is for functions that consume themselves, such as `FnOnce(self)`. They can be called only once because they consume the variables they capture.

In the case of closures, `FnOnce` is always implemented if the closure consumes any of the variables it captures, denoted by the `move` keyword before the definition of a closure. Consider the closure in the following listing.

Listing 3.3 Closure with move

```
let consumable = String::from("cookie");
let consumer = move || consumable;
consumer();
// consumer(); error!
```

In this example, the fourth line would produce an error because our `consumable` can be moved only once, so calling `consumer()` a second time is invalid. If we try compiling with the second call to `consumer()` uncommented, we'll get the following output from the compiler:

```
error[E0382]: use of moved value: `consumer`
  --> src/main.rs:22:5
   |
21 |     consumer();
   |     ---------- `consumer` moved due to this call
22 |     consumer();
   |     ^^^^^^^^^^ value used here after move
   |
note: closure cannot be invoked more than once because it moves the
⇒ variable `consumable` out of its environment
  --> src/main.rs:20:28
   |
20 |     let consumer = move || consumable;
   |                            ^^^^^^^^^^
note: this value implements `FnOnce`, which causes it to be moved when
⇒ called
  --> src/main.rs:21:5
```

```
   |
21 |       consumer();
   |       ^^^^^^^^
```

For more information about this error, try `rustc --explain E0382`.
error: could not compile `closures` (bin "closures") due to 1 previous
➥ error

The primary use of `move |...|` (as in listing 3.3) is when you want to transfer or assign ownership of an object somewhere inside the closure but avoid copying or cloning it. The `move` keyword is optional; if you don't use it, Rust infers whether to move the variables you capture. Still, being explicit about your intentions is a good idea because it prevents ambiguity. The compiler will alert you if an error occurs, of course. In the example with `consumable`, we could have omitted the `move` keyword safely; the result would have been the same. We can combine the use of closures, generics, and the `Fn`, `FnMut`, and `FnOnce` traits to enable a variety of generic functional patterns.

3.2.3 *Examining iterators*

Let's take a look at Rust's iterators, which complement closures. Rust's iterators are provided by the `Iterator` trait, which includes a lot of functionality built on top of iterators: `map()`, `for_each()`, `take()`, `fold()`, `filter()` `find()`, `zip()`, and more. If you implement the `Iterator` trait for your type, you receive all these iterators (and more!).

Iterators are one of the original Gang of Four design patterns and arguably the most prolific. They provide a great case study not only for design patterns but also for the Rust language. The core of Rust's `Iterator` trait is as follows:

```
trait Iterator {
    type Item;
    fn next(&mut self) -> Option<Self::Item>;
}
```

The `Iterator` trait contains a lot more than what you see here, but if you want to implement `Iterator` for your type, you need to provide only `next()` and `Item`. Let's examine an example of a linked list in Rust by implementing the `Iterator` trait. We'll start by writing a new linked list implementation.

Listing 3.4 Implementing `LinkedList`

```
use std::cell::RefCell;
use std::rc::Rc;

type ItemData<T> = Rc<RefCell<T>>;
type ListItemPtr<T> = Rc<RefCell<ListItem<T>>>;

struct ListItem<T> {
    data: ItemData<T>,                ⟵────┘ A pointer to
                                              our data
    next: Option<ListItemPtr<T>>,     ⟵──┐ A pointer to the next
}                                          item in the linked list
```

```
impl<T> ListItem<T> {
    fn new(t: T) -> Self {                    ◄─┤  Creates a new item
        Self {                                     (or node) for the list
            data: Rc::new(RefCell::new(t)),
            next: None,
        }
    }
}

struct LinkedList<T> {                        ┐  A pointer to the first item
    head: ListItemPtr<T>,          ◄─┘  (or node) in the list
}

impl<T> LinkedList<T> {                       ┐  Creates a new list, with the
    fn new(t: T) -> Self {         ◄─┘  head pointing to the first item
        Self {
            head: Rc::new(RefCell::new(ListItem::new(t))),
        }
    }
}
```

We have an incomplete linked list that has the structure we need but doesn't give us a way to iterate over the list or append new items. I intentionally left out the append functionality because I want to use an iterator to implement it. If I implement `Iterator` first, the rest of the linked list features become easy to add. Let's give it a shot.

Rc and RefCell

If you haven't encountered `Rc` or `RefCell` (introduced in listing 3.4), don't panic; I'll provide a brief explanation for readers who aren't familiar with them. In short, `Rc` and `RefCell` are *smart pointers* that provide important (but distinct) features.

`Rc` provides a reference-counted pointer, similar to C++'s `std::shared_ptr`. `RefCell` is a special type of pointer that enables *interior mutability*.

`Rc` allows you to hold multiple references (or pointers) to the same location in memory, and `RefCell` provides a way to perform borrow checking at run time. Rust's borrow checker normally works at compile time, but sometimes you want to perform the borrow checking at run time instead, such as when you want to hold multiple references to the same object and still enable mutability (not possible at compile time).

In our linked list example, we need to hold multiple references to the same object (which `Rc` provides), and we also want to be able to mutate the inner object (which `RefCell` allows us to do safely).

In chapter 5 of *Code Like a Pro in Rust* (https://www.manning.com/books/code-like -a-pro-in-rust), I discuss Rust's smart pointers at great length. For details on `Rc`, consult the Rust standard library documentation at https://doc.rust-lang.org/std/rc/ index.html, and for `RefCell`, refer to https://doc.rust-lang.org/std/cell/index.html.

I'll note here that iterators are *stateful*. That is, an iterator knows where it is in the sequence of items so that it can go from the previous to the next item with each subsequent call to next().

> **NOTE** Even in the purest functional programming languages, you can always find state under the hood if you look hard enough, as all software eventually breaks down to strictly imperative machine code.

For now, we'll store that state in our linked list itself. We can update the structure this way, along with the fn new() method:

```
struct LinkedList<T> {
    head: ListItemPtr<T>,
    cur_iter: Option<ListItemPtr<T>>,
}

impl<T> LinkedList<T> {
    fn new(t: T) -> Self {
        Self {
            head: Rc::new(RefCell::new(ListItem::new(t))),
            cur_iter: None,
        }
    }
}
```

Great! Now we have a pointer to the current position of our iterator in cur_iter, which can be initialized to None. Let's take a first shot at implementing the Iterator trait for our linked list (not the refined approach, which we'll arrive at later in this chapter):

For this Iterator implementation, we'll return a pointer to the list item rather than the data itself.

```
impl<T> Iterator for LinkedList<T> {
    type Item = ListItemPtr<T>;
    fn next(&mut self) -> Option<Self::Item> {
        match &self.cur_iter.clone() {
            None => {
                self.cur_iter = Some(self.head.clone());
            }
            Some(ptr) => {
                self.cur_iter = ptr.borrow().next.clone();
            }
        }
        self.cur_iter.clone()
    }
}
```

We have to clone cur_iter here because we try to modify the pointer while it's borrowed later.

If cur_iter is None, the iterator is uninitialized, so we start at the head.

cur_iter must be updated to point to the next item in the sequence.

Last, we clone and return the current position in our sequence.

Now finding the last item in the list with an iterator is a trivial operation:

```
let dinosaurs = LinkedList::new("Tyrannosaurus Rex");
let last_item = dinosaurs.last()
```

```
    .expect("couldn't get the last item");
println!("last_item='{}'", last_item.borrow().data.borrow());
```

By implementing `Iterator`, we can call `last()` to retrieve the last item in our list, which we get for free from the `Iterator` trait. Running the preceding code prints `last_item= 'Tyrannosaurus Rex'`, as we'd expect. Now let's add our `append()` method to the original `LinkedList`:

```
impl<T> LinkedList<T> {
    fn new(t: T) -> Self {
        Self {
            head: Rc::new(RefCell::new(ListItem::new(t))),
            cur_iter: None,
        }
    }
    fn append(&mut self, t: T) {
        self.last()
            .expect("List was empty, but it should never be")      We must borrow
            .as_ref()                                               the inner RefCell
            .borrow_mut()                                    ◄─     to access the inner
            .next = Some(Rc::new(RefCell::new(ListItem::new(t)))); ◄ ListItem.
    }
}                                       We have to borrow mutably to
                                        modify the inner next pointer.
```

Now we can append and then iterate over our list by using `for_each` with a closure:

```
let mut dinosaurs = LinkedList::new("Tyrannosaurus Rex");
dinosaurs.append("Triceratops");
dinosaurs.append("Velociraptor");
dinosaurs.append("Stegosaurus");
dinosaurs.append("Spinosaurus");
dinosaurs                                              We still have to
    .iter()                                            unwrap the inner
    .for_each(|ptr| {                                  pointer here, and our
                                                       call to for_each() will
      println!("data={}", ptr.borrow().data.borrow()) ◄ consume dinosaurs.
    );
```

Running this code prints the following:

```
data=Tyrannosaurus Rex
data=Triceratops
data=Velociraptor
data=Stegosaurus
data=Spinosaurus
```

> **NOTE** The code in this example doesn't match the final implementation and, therefore, doesn't match the code in the repository, but we'll get there soon.

Neat, huh? This example is fun, but our iterator is less than ideal because we still have to unwrap the internal pointer to access our payload data within each node of the linked list. In my opinion, this interface is pretty awkward for a collection type. We probably wouldn't want to expose our internal types if we were writing a library.

3.2.4 *Obtaining an iterator with iter(), into_iter(), and iter_mut()*

To make our linked list more idiomatic, we need to iterate over items in the list without exposing the internal structure of the list. We also need to iterate over mutable references to the items in the list and to consume the list and iterate over the items. In other words, we may want to iterate over our linked list in three ways:

- `iter()`—Iterate over immutable references to the items in the list.
- `iter_mut()`—Iterate over mutable references to the items in the list.
- `into_iter()`—Consume the list and iterate over the items.

In section 3.2.3, I implemented the `Iterator` trait directly on `LinkedList`, but this is not idiomatic Rust, and it's bad practice. Instead, we'll create a separate structure to handle iteration, which is a common pattern in Rust and better design. If we look at Rust's built-in collection types, they typically provide three iterators:

- An iterator that iterates over `T`, provided by `into_iter(self)`, which consumes `self`
- An iterator that iterates over `&T`, provided by `iter(&self)`
- An iterator that iterates over `&mut T`, provided by `iter_mut(&mut self)`

You'll notice that `Vec` does not implement the `Iterator` trait directly; instead, it implements the `IntoIterator` trait for `T`, `&T`, and `&mut T`. `Vec` uses its own internal (https://doc.rust-lang.org/std/vec/struct.IntoIter.html) `Iter`, `IterMut`, and `IntoIter` objects to implement the `Iterator` trait instead of doing it directly on `Vec`. We can do the same with our linked list by creating separate structures to handle iteration rather than implementing `Iterator` for `LinkedList`.

Let's copy this pattern and apply it to our linked list. First, we'll create our new stateful iterator structs, which look like this:

```
struct Iter<T> {
    next: Option<ListItemPtr<T>>,
}
struct IterMut<T> {
    next: Option<ListItemPtr<T>>,
}
struct IntoIter<T> {
    next: Option<ListItemPtr<T>>,
}
```

Each iterator struct maintains a pointer to the next item in the list. Because we're using `Rc` and `RefCell` to implement the linked list, managing the pointers is fairly easy, and we don't have to worry much about lifetimes.

We'll initialize these iterators by adding `iter()`, `iter_mut()`, and `into_iter()` methods to `LinkedList`, which returns a new instance. We'll also update our `append()` so that it works again:

```
impl<T> LinkedList<T> {
    fn new(t: T) -> Self {
        Self {
            head: Rc::new(RefCell::new(ListItem::new(t))),
        }
    }
    fn append(&mut self, t: T) {
        let mut next = self.head.clone();
        while next.as_ref().borrow().next.is_some() {
            let n = next
                .as_ref()
                .borrow()
                .next
                .as_ref()
                .unwrap()
                .clone();
            next = n;
        }
        next.as_ref().borrow_mut().next =
            Some(Rc::new(RefCell::new(ListItem::new(t))));
    }
    fn iter(&self) -> Iter<T> {
        Iter {
            next: Some(self.head.clone()),
        }
    }
    fn iter_mut(&mut self) -> IterMut<T> {
        IterMut {
            next: Some(self.head.clone()),
        }
    }
    fn into_iter(self) -> IntoIter<T> {
        IntoIter {
            next: Some(self.head.clone()),
        }
    }
}
```

We have to unwrap the inner Option within the RefCell and Rc, which is why we need to obtain a reference with as_ref() and borrow with borrow() to access the inner next pointer.

We have to borrow three times: twice from the current next and once from the next next, after which we can clone the pointer.

Cool! We've updated `append()` so that it no longer uses the old `Iterator` implementation, which we've already decided is flawed. Now all we have to do is implement the `Iterator` trait for `Iter`, `IterMut`, and `IntoIter`:

```
impl<T> Iterator for Iter<T> {
    type Item = ItemData<T>;
    fn next(&mut self) -> Option<Self::Item> {
        match self.next.clone() {
            Some(ptr) => {
                self.next.clone_from(&ptr.as_ref().borrow().next);
                Some(ptr.as_ref().borrow().data.clone())
```

```
            }
            None => None,
        }
    }
}
impl<T> Iterator for IterMut<T> {
    type Item = ItemData<T>;
    fn next(&mut self) -> Option<Self::Item> {
        match self.next.clone() {
            Some(ptr) => {
                self.next.clone_from(&ptr.as_ref().borrow().next);
                Some(ptr.as_ref().borrow().data.clone())
            }
            None => None,
        }
    }
}
impl<T> Iterator for IntoIter<T> {
    type Item = ItemData<T>;
    fn next(&mut self) -> Option<Self::Item> {
        match self.next.clone() {
            Some(ptr) => {
                self.next.clone_from(&ptr.as_ref().borrow().next);
                Some(ptr.as_ref().borrow().data.clone())
            }
            None => None,
        }
    }
}
```

Our `next()` implementation is straightforward: we return the pointer to the data within our `ListItem` struct, update `self.next` to the next item in the list, and return `None` when there are no more entries. You may notice that all three implementations are identical. The situation is even worse: all of them return `Rc<RefCell<T>>` rather than the `T`, `&T`, and `&mut T` we're looking for. Returning `Rc<RefCell<T>>` is fine, but it doesn't match the pattern, and we still have to unwrap the data to access it.

The solution to this problem isn't straightforward, but let's try to fix it by looking at `IntoIter` from `Vec`. The `into_iter()` method on `Vec` has the following signature:

```
fn into_iter(self) -> slice::IterMut<'a, T>;
```

If you look carefully, you'll see that the method takes `self` by value. In other words, calling `into_iter()` consumes the `Vec`. We can use this knowledge to change our `IntoIter` so that it consumes each list item:

```
impl<T> Iterator for IntoIter<T> {
    type Item = T;
    fn next(&mut self) -> Option<Self::Item> {
        match self.next.clone() {
            Some(ptr) => {
                self.next = ptr.as_ref().borrow().next.clone();
```

```
        let listitem =
            Rc::try_unwrap(ptr).map(|refcell| refcell.into_inner());
        match listitem {
            Ok(listitem) => Rc::try_unwrap(listitem.data)
                .map(|refcell| refcell.into_inner())
                .ok(),
            Err(_) => None,
        }
    }
    None => None,
    }
}
}
```

The code is starting to look a lot more complicated. Let's break it down:

- Both our pointers to each list item (or node) in the linked list, as well as the data, are stored in a `RefCell` inside `Rc` (i.e., `Rc<RefCell<T>>`).
- We need to use `try_unwrap()` on the `Rc` to move the inner `RefCell` out of the `Rc` because we want to consume it. `try_unwrap()` works on `Rc` only when there are no other references. Because we're not going to expose these references outside our linked list, we can be reasonably sure that there aren't any other references.
- When we get the `RefCell` out of the `Rc` using `try_unwrap()`, we need to move the `T` out of `RefCell<T>`. To do so, we call `into_inner()`, which consumes the `RefCell` that returns an owned `T`.
- The return type is defined by `type Item = T`, which is an associated type, and we reference it with `Self::Item`, which is required by the `Iterator` trait.

We can test our code this way:

```
let mut dinosaurs = LinkedList::new("Tyrannosaurus Rex");
dinosaurs.append("Triceratops");
dinosaurs.append("Velociraptor");
dinosaurs.append("Stegosaurus");
dinosaurs.append("Spinosaurus");
dinosaurs
    .into_iter()
    .for_each(|data| println!("data={}", data));
```

The test works as expected, producing the following output:

```
data=Tyrannosaurus Rex
data=Triceratops
data=Velociraptor
data=Stegosaurus
data=Spinosaurus
```

Neat! Let's look at our `Iter` and `IterMut` implementations again because they still don't return `&T` or `&mut T` the way we want. Unlike `into_iter()`, the `iter()` and `iter_mut()`

methods on `LinkedList` don't consume `self`; they take references to `self` (`&self` and `&mut self`, respectively), which makes things quite tricky.

In stable Rust, `RefCell` doesn't provide a way to get a plain reference to the object it holds. The `Ref` and `RefMut` wrappers provide a `leak()` method in Rust nightly, but let's try to do it without using that feature.

Unfortunately, the only way to do what we want is to use `unsafe`. If you look at Rust's collection library implementations, you'll see that they use `unsafe` in various places, such as the internal implementation of `next()` from the `Iterator` trait.

We need to update the `Iter` and `IterMut` structs to include a lifetime `'a` for the reference we're returning. We'll also store a copy of the pointer to the data we're returning so that it exists as long as the iterator is in scope. We use a `PhantomData` field to capture the lifetime `'a` in the struct:

```
struct Iter<'a, T> {
    next: Option<ListItemPtr<T>>,
    data: Option<ItemData<T>>,
    phantom: PhantomData<&'a T>,
}
struct IterMut<'a, T> {
    next: Option<ListItemPtr<T>>,
    data: Option<ItemData<T>>,
    phantom: PhantomData<&'a T>,
}
```

Lifetimes

Lifetimes ensure that references are valid for a certain period to prevent dangling references (akin to dangling pointers in C or C++). Rust introduced the concept of lifetimes to allow the compiler's borrow checker to verify that references are valid at compile time and give programmers a way to communicate this information to the compiler. Lifetimes are denoted by an apostrophe (`'`) followed by a name, such as `'a`, `'b`, and `'c`.

Rust's lifetimes are a bit tricky to grok at first, but with practice, you'll see that they're quite simple. Here are a few important points to consider regarding lifetimes:

- A variable's lifetime is the period for which it's valid, beginning when the variable is created and ending when it is destroyed.
- A reference is valid for the lifetime `'a`, where `'a` is an arbitrary name that carries no meaning other than to identify the lifetime.
- A reference is valid for the lifetime of the object it references or the lifetime of the scope in which it was created, whichever is shorter.
- Sometimes, we have to define lifetimes explicitly to help the compiler understand the relationship between references. At other times, the compiler can infer the lifetimes for us (generally the default).
- If the compiler can't infer the lifetimes, it produces an error message, and you'll need to provide the lifetimes explicitly.

- Lifetimes always exist in the context of a reference and are always associated with a reference. You don't need a lifetime if you don't have a reference, and the compiler will infer a lifetime for you if you don't define one explicitly.

Lifetimes are generally introduced at the function, struct, or trait level. Where the lifetime is introduced determines the scope of the lifetime. If you introduce a lifetime at the function level, the lifetime is valid for the duration of the function (or struct, trait, or so on). Consider the following small program, which introduces the functions `print_without_lifetime()` and `print_with_lifetime()`:

```
fn print_without_lifetime(s: &str) {
    println!("{}", s);
}

fn print_with_lifetime<'a>(s: &'a str) {
    println!("{}", s);
}

fn main() {
    print_without_lifetime("calling print_without_lifetime()");
    print_with_lifetime("calling print_with_lifetime()");
}
```

The two functions are identical except that `print_with_lifetime()` has an explicit lifetime `'a` defined for the reference to the string `s`. The compiler will infer the lifetime for `print_without_lifetime()`, but we explicitly define the lifetime for `print_with_lifetime()`.

Adding the lifetime `'a` to the function signature tells the compiler that the reference is valid for the duration of the function, which in this case is simply the duration of the function call.

If you were to add a lifetime to the definition of a struct instead, the lifetime would be valid for the duration of the struct object. Consider the following example:

```
struct RefStruct<'a> {
    s_ref: &'a str,
}

fn main() {
    let dog = "dog";
    let dog_struct = RefStruct { s_ref: dog };    ⟵── dog_struct must
    println!("I am a {}", dog_struct.s_ref)            not outlive dog.
}
```

In this code, the lifetime `'a` is introduced at the struct level, which means that the reference `s_ref` is valid for the duration of the struct `RefStruct`. Now we can put a reference to `dog` in the struct `RefStruct` and print it as long as `dog` outlives `dog_struct`.

If this concept doesn't make complete sense just yet, don't worry; it will become more apparent as you spend more time with Rust. For more information on lifetimes, see the section on lifetimes at https://mng.bz/QZ91.

We also need to initialize the new `data` and `phantom` fields in `iter()` and `iter_mut()`:

```
impl<T> LinkedList<T> {
    fn iter(&self) -> Iter<T> {
        Iter {
            next: Some(self.head.clone()),
            data: None,
            phantom: PhantomData,
        }
    }
    fn iter_mut(&mut self) -> IterMut<T> {
        IterMut {
            next: Some(self.head.clone()),
            data: None,
            phantom: PhantomData,
        }
    }
}
```

Now we can implement the `next()` method for both:

```
impl<'a, T> Iterator for Iter<'a, T> {
    type Item = &'a T;
    fn next(&mut self) -> Option<Self::Item> {
        match self.next.clone() {
            Some(ptr) => {
                self.next = ptr.as_ref().borrow().next.clone();
                self.data = Some(ptr.as_ref().borrow().data.clone());
                unsafe { Some(&*self.data.as_ref().unwrap().as_ptr()) }
            }
            None => None,
        }
    }
}
impl<'a, T> Iterator for IterMut<'a, T> {
    type Item = &'a mut T;
    fn next(&mut self) -> Option<Self::Item> {
        match self.next.clone() {
            Some(ptr) => {
                self.next = ptr.as_ref().borrow().next.clone();
                self.data = Some(ptr.as_ref().borrow().data.clone());
                unsafe { Some(&mut *self.data.as_ref().unwrap().as_ptr()) }
            }
            None => None,
        }
    }
}
```

As you can see, we've got to do some pointer coercion to get what we want. We use the `as_ptr()` method on `RefCell` to get `*mut T`; next, we dereference that pointer; then we take another reference. This approach isn't pretty, but it works. Keep in mind that this structure isn't thread-safe. Finally, we can test it, and the code prints what we expect:

```
let mut dinosaurs = LinkedList::new("Tyrannosaurus Rex");
dinosaurs.append("Triceratops");
dinosaurs.append("Velociraptor");
dinosaurs.append("Stegosaurus");
dinosaurs.append("Spinosaurus");
dinosaurs
    .iter()
    .for_each(|data| println!("data={}", data));

dinosaurs
    .iter_mut()
    .for_each(|data| println!("data={}", data));
```

One more thing: we need to add the `IntoIterator` trait and remove the previous `impl<T> Iterator for LinkedList<T> {}` block. By doing so, we can iterate over our list by using a `for` loop:

```
impl<'a, T> IntoIterator for &'a LinkedList<T> {
    type IntoIter = Iter<'a, T>;
    type Item = &'a T;
    fn into_iter(self) -> Self::IntoIter {
        self.iter()                          ◁────── Wraps iter()
    }                                               on LinkedList
}
impl<'a, T> IntoIterator for &'a mut LinkedList<T> {
    type IntoIter = IterMut<'a, T>;
    type Item = &'a mut T;
    fn into_iter(self) -> Self::IntoIter {       Wraps iter_mut()
        self.iter_mut()                          on LinkedList
    }                             ◁──────
}                                                We don't need the 'a
impl<T> IntoIterator for LinkedList<T> {  ◁────── lifetime parameter here
    type IntoIter = IntoIter<T>;                 because it's not used later.
    type Item = T;
    fn into_iter(self) -> Self::IntoIter {
        self.into_iter()                  ◁──────
    }                                                Wraps into_iter()
}                                                    on LinkedList
```

We can test the code as follows, using a plain old `for` loop:

```
for data in &linked_list {
    println!("with for loop: data={}", data);
}
```

The compiler knows which implementation of `IntoIterator` to use based on the type passed to the `for` loop. In this case, we're passing `&linked_list`, so the compiler uses the form that returns `&T`, calling the `iter()` method on `LinkedList`.

When you have iterators implemented, they unlock a lot of built-in functionality, including `for_each()`, `map()`, `reduce()`, `filter()`, `zip()`, and `fold()`. You can also use `for … {}` with structures that implement `IntoIterator` or `Iterator`.

NOTE I generally prefer using the `for_each()` method as opposed to the `for ... {}` loop syntax, although these approaches are functionally equivalent. `for_each()` accepts a function as its argument, which means that you can pass a closure or another function to it directly. In special cases, such as when you're using `async`/`await`, you must use a `for` loop rather than `for_each()`.

3.2.5   *Iterator features*

Let's take a quick tour of the features that iterators unlock. Here's an example of `map()`:

```
let arr = [1, 2, 3, 4];
println!("{:?}", arr);
let vec: Vec<_> = arr.iter().map(|v| v.to_string()).collect();
println!("{:?}", vec);
```

First, we initialize an array with some integers. Next, we convert our integers to strings of integers (that is, print them to a string). To do that, we map each value to a string by using `map()`. `map()` takes a function as its argument; it's a higher-order function. Let's take a quick look at the signature of `map()`:

```
fn map<B, F>(self, f: F) -> Map<Self, F>
where
    F: FnMut(Self::Item) -> B,
{ ... }
```

The `map()` method takes a function with one parameter, `Self::Item`, as noted by the trait bounds. If you recall from the `Iterator` trait, `Self::Item` is defined by the iterator itself. In the case of a slice, array, or `Vec`, `Self::Item` is `&T`. That function can return any type, denoted by the `B` generic parameter. What's most interesting about `map()` is that it merely returns another iterator, this time a special one called `Map` that Rust provides. We pass a closure to `map()`, but we could also supply the `i32::to_string()` function directly as an argument.

TIP Rust's iterators use lazy evaluation when possible, such as with `map()`. The results are not computed until you force evaluation (such as by calling `collect()`).

The last step is calling `collect()`, which converts an iterator to a collection—usually, a `Vec`. You'll notice that we have to tell the compiler what the target type is because it can't figure out the type automatically. Running the preceding code produces the following output:

```
[1, 2, 3, 4]
["1", "2", "3", "4"]
```

Suppose that we want to do something slightly more elaborate. Perhaps we want to convert a `Vec` to a `LinkedList` from the Rust standard library while also applying a transformation. Let's reuse the second `vec` from the preceding example and parse our strings back into integers:

```
let linkedlist: LinkedList<i32> =
    vec.iter().flat_map(|v| v.parse::<i32>()).collect();
println!("{:?}", linkedlist);
```

We did something new by using `flat_map()` instead of `map()`. Why are we using `flat_map()`? Because `String::parse()` returns a `Result`, so we need to flatten the result of that parsing operation. We could call `unwrap()` after parsing, but `flat_map()` is a little cleaner, and it handles errors somewhat gracefully (by tossing them aside).

To elaborate, `flat_map()` flattens the `Result` by calling the `Result::into_iter()` method, which returns an iterator over the `Ok` value if it's present or an empty iterator if it's not. The `Err` value is ignored when the `Result` is flattened.

The problem is that if our parsing contains an error, we might not catch it. Not to worry. `partition()` has our back:

```
let arr = ["duck", "1", "2", "goose", "3", "4"];
let (successes, failures): (Vec<_>, Vec<_>) = arr
    .iter()
    .map(|v| v.parse::<i32>())
    .partition(Result::is_ok);
println!("successses={:?}", successes);
println!("failures={:?}", failures);
```

Here, we're taking a list of strings and trying to parse each string into an integer. Because we managed to get a `duck` and a `goose` in there (they aren't integers), parsing them will fail. We want to split, or *partition*, the result of the parsing job, so we're going to partition on `Result::is_ok()`, which returns `true` if the result is `Ok`. Running the preceding code prints the following:

```
successses=[Ok(1), Ok(2), Ok(3), Ok(4)]
failures=[Err(ParseIntError { kind: InvalidDigit }),
Err(ParseIntError { kind: InvalidDigit })]
```

That's odd—our successes and failures are still wrapped in a `Result`, which makes sense because we didn't unwrap them. We can unwrap them with another step:

```
let successes: Vec<_> =
    successes.into_iter().map(Result::unwrap).collect();
let failures: Vec<_> =
    failures.into_iter().map(Result::unwrap_err).collect();
println!("successses={:?}", successes);
println!("failures={:?}", failures);
```

Notice that we're calling `into_iter()` on our `Vec` because when we unwrap the `Result`, we also want to consume it. `into_iter()`, if you recall, consumes the `Vec` and its contents. Running the preceding code produces the following:

```
successses=[1, 2, 3, 4]
failures=[ParseIntError { kind: InvalidDigit },
ParseIntError { kind: InvalidDigit }]
```

Sweet! Everything is as it should be.

> **TIP** Try to avoid using constructs such as for and while loops; instead, use collections with iterators. Instead of a for loop, you can use for_each(), and instead of a while loop, you can use map_while().

We can get quite elaborate in chaining operations with iterators. Rust also provides a few special-purpose iterators to handle more complex tasks, such as counting with Enumerate. Here's an example that shows how we might use Enumerate with a list of dog breeds:

```
let popular_dog_breeds = vec![
    "Labrador",
    "French Bulldog",
    "Golden Retriever",
    "German Shepherd",
    "Poodle",
    "Bulldog",
    "Beagle",
    "Rottweiler",
    "Pointer",
    "Dachshund",
];

let ranked_breeds: Vec<_> =
    popular_dog_breeds.into_iter().enumerate().collect();

println!("{:?}", ranked_breeds);
```

Running this code yields the following output:

```
[(0, "Labrador"), (1, "French Bulldog"), (2, "Golden Retriever"),
(3, "German Shepherd"), (4, "Poodle"), (5, "Bulldog"), (6, "Beagle"),
(7, "Rottweiler"), (8, "Pointer"), (9, "Dachshund")]
```

That's close but probably not quite what we want. It would make sense to start the count at 1 instead of 0. With a small change, we can improve the code to get the result we're looking for:

```
let ranked_breeds: Vec<_> = popular_dog_breeds
    .into_iter()
    .enumerate()
    .map(|(idx, breed)| (idx + 1, breed))
    .collect();
```

We added a map() after enumerate() to unpack the tuple produced by enumerate() and return it with 1 added to the index. Now we get the result we want:

```
[(1, "Labrador"), (2, "French Bulldog"), (3, "Golden Retriever"),
(4, "German Shepherd"), (5, "Poodle"), (6, "Bulldog"), (7, "Beagle"),
(8, "Rottweiler"), (9, "Pointer"), (10, "Dachshund")]
```

What if we want to count down instead of up? We can reverse the list with `rev()`:

```
let ranked_breeds: Vec<_> = popular_dog_breeds
    .into_iter()
    .enumerate()
    .map(|(idx, breed)| (idx + 1, breed))
    .rev()
    .collect();
```

Iterators are among my favorite abstractions in Rust. It's remarkable how quickly you can go from a quick-and-dirty data structure to a full-featured collection simply by implementing a few iterator traits.

> **TIP** For a complete list of all features provided by Rust's iterators, consult the standard library reference at https://doc.rust-lang.org/std/iter/index.html.

Between iterators and closures, Rust provides what you need to write purely functional code easily. Rust's memory model does make it trickier to perform specific tasks in Rust that may be trivial in other languages, but almost no other language can compete with Rust in terms of features, safety, and performance.

Summary

- Pattern matching allows us to unpack data structures and handle a variety of scenarios in a much cleaner way than using combinations of `if`/`else` statements.
- We can use pattern matching with the `?` operator to handle errors gracefully and unwrap or destructure values.
- We can destructure nested structs and enums when pattern matching, and we can also match on values.
- Rust encourages functional programming patterns, particularly with closures and iterators. Learning these patterns will help you use Rust effectively.
- Iterators use a fluent interface, and along with closures, we can easily express operations and mutations on data structures.
- Iterators typically hold a reference to the data (such as borrowed data) or use a move to move the items out of the underlying sequence.
- Usually, the `iter()` method returns an iterator with references, and `into_iter()` gives us an iterator that takes ownership with a move.

Part 2

Core patterns

Core patterns are those that we use over and over, almost to the point at which they become clichés. So it's vital for our success to master these core patterns and understand them well. Also, we need to make sure that we speak the language of patterns in a way that enables us to communicate our systems and designs to other people.

Occasionally, it's good to remember that patterns are not the goal. We may need to step back from our work and view it from a higher level to ensure that we're not applying patterns mindlessly without understanding the problem we're trying to solve.

The goal of software design is rarely to use all the features of a language or maximize the number of lines of code. Rather, the goal is to solve problems and create enduring value. Patterns are tools that help us reach that goal, but they're not the only ones we have at our disposal. Sometimes, the best solution is the simplest one or the one that is best understood. We also write software for fun sometimes, and that's fine too.

Introductory patterns

This chapter covers

- Understanding resource acquisition is initialization
- Passing arguments by value versus reference
- Using constructors
- Understanding object member visibility and access
- Handling errors
- Global state handling with `lazy-static.rs`, `OnceCell`, and `static_init`

Now we're ready to dive into some more concrete patterns. We begin by reviewing some elementary topics: RAII, passing values, constructors, and visibility. Then we'll move on to slightly more complex subjects: error handling and global variables. Although the chapter discusses many topics, it focuses on bite-size patterns, which we'll use a lot.

This chapter also introduces *crates*, which are Rust libraries built by the community. The Rust language is built on crates, which are crucial parts of Rust programming; you won't get far without using them. Although it's possible to go full not-invented-here syndrome and eschew crates, I don't recommend this approach. Even the largest, best-funded organizations rely heavily on open source software to build their stacks to varying degrees.

You'll quickly find when working with Rust that the standard library is somewhat bare and doesn't include many of the features you might expect from a modern language. These limits are by design; the Rust team chose to keep the standard library minimal and instead rely on crates to provide additional functionality. This approach has several benefits:

- The standard library is smaller and easier to maintain.
- The standard library is more stable and less likely to change.
- The standard library is more focused on core functionality.
- The community can build and maintain separate competing crates for specialized functionality, allowing developers to choose the most suitable crate for their needs.

If you want to work exclusively with proprietary software, you should pay attention to the licenses provided by each crate. Because this book is intended to be educational, I will assume that you're fine with relying on open source software with licenses that may not be compatible with commercial or proprietary use. The bulk of Rust crates use permissive licenses, which permit nearly any use.

4.1   *Resource acquisition is initialization*

Resource acquisition is initialization (typically referred to as RAII) originated with C++ and is arguably one of the most important modern programming idioms. RAII is a key feature in Rust: it allows us to confidently implement a variety of other patterns and plays a critical role in Rust's safety features.

There's some question about whether RAII is an idiom or a pattern, but I'll describe it as a pattern rather than an idiom because it's a formalized way of handling resources in a program, as opposed to a more informal way of formatting code. Additionally, RAII affects the overall program structure and architecture, which is more in line with a pattern than an idiom.

4.1.1   *Understanding RAII in C and C++*

In this section, I'll quickly explain RAII and how it works in case you've never encountered the concept. For any seasoned programmer, this section is likely to be a review of a well-understood concept. We'll examine some C and C++ code because C++ gave birth to RAII as an improvement to C. If you're unfamiliar with either language, don't worry; the examples are simple, and you don't need to understand them in depth.

RAII uses the stack within a particular scope to determine when resources (such as variables) can be released. The name may be confusing because RAII is usually thought of as a way to handle the release of resources instead of the acquisition and initialization of resources, as the name implies. These functions are related, however, so let me explain further. To begin, let's examine what happens if we declare a simple variable within a function in C:

```
void func() {
    int a;
    // Some code goes here that does something with a.
}
```

In this C function, we declare a variable a. Although we've *declared* the variable in our function, we haven't *initialized* it, which we do by assigning a value to the variable. Thus, the value of a in the example is undefined because it hasn't been initialized. Commonly, you'll see code like the following snippet in C, which handles both the declaration and initialization:

```
void func() {
    int a = 0;
}
```

This code declares *and* initializes a to the value of 0. We know now that a is 0 at the time of declaration. When the function returns, a goes out of scope and is popped off the stack, which means the variable is released. The C language doesn't do anything special when a variable is released.

Now, what happens when a is a pointer? In other words, if a points to memory somewhere else, what happens when a is released? In C, we might have some code like this:

```
void func() {
    int *a = malloc(sizeof(int));
}
```

This code creates a memory leak because we're allocating memory from the heap with `malloc()` and assigning the address to a, which is returned by the `malloc()` function. Note that `sizeof(int)` contains the size in bytes of an `int` or integer, which is often 4 bytes, but this setting is platform-dependent.

When we return from this function, the pointer a is released, but the memory blocks that our pointer addresses are not released, so we've created a memory leak. The solution in this case is to call `free(a)` to release the address at a before returning from the function.

But here's the problem: What if we can return from multiple places within our function? Suppose that we write the following code:

```
void leaky_func() {
    FILE *fp;
    int *a = malloc(sizeof(int));        Initializes the
    *a = 0;                          ◁── value of a to 0
```

```
// try to open a file for reading
fp = fopen("file.txt", "r");
if (fp == NULL) {
    // there was an error!
    return;
}

// Now we can read from the file at fp.
// ...

fclose(fp);                              Closes the
                                         file pointer
free(a);
                                         Releases the memory
}                                        pointed to by a
```

The function `leaky_func` opens a file for reading, but if a failure occurs when opening the file (such as when the file doesn't exist), we return from our function early. We've also introduced a memory leak because we won't release the memory from a when a failure occurs. This situation is a classic memory leak and one of the downsides of working with languages like C.

One of C++'s ambitions was to make it harder to introduce memory leaks, and one way it did so was by using *constructors* and *destructors*. When you create a class or struct in C++, it always calls the constructor at the time of creation. When you destroy an object in C++, it always calls the destructor. If you create an object on the stack in C++, it automatically calls the constructor and destructor for you. But if you create an object on the heap, you need to use the `new` and `delete` keywords to release the memory and call the constructors and destructors, respectively. `new` and `delete` in C++ are equivalent to `malloc()` and `free()` in C. These keywords don't solve the memory leak problem, but RAII helps you avoid memory leaks by using smart pointers.

A *smart pointer* is a special kind of pointer that provides a constructor that wraps `new` and a destructor that wraps `delete`. Because the compiler guarantees that any variable going out of scope will have its destructor called, we can build on top of this behavior to effectively eliminate one class of memory leaks, but only if we always use smart pointers.

To make matters more complicated, C++ is backward-compatible with C, so C code is perfectly valid C++. For this reason, C++ provides as much opportunity to shoot yourself in the foot as C does despite the introduction of constructors, destructors, and smart pointers.

As you can probably guess, although C++ gave people the tools to solve one class of memory leaks, they didn't always use the tools correctly (or at all), so C++ made only small strides in fixing this problem. The C++ equivalent to the preceding C code, this time using `std::shared_ptr` instead of a plain C pointer, looks something like this:

```
#include <fstream>
#include <memory>

void func() {
    std::shared_ptr<int> a(new int(0));
```

```
    std::ifstream stream("file.txt");
    if (!stream.is_open()) {
        // error!
        return;
    }

    // Now we can read from our file.
    // ...
}
```

Notice that we use `std::shared_ptr` for our pointer a, which eliminates the memory leak. It no longer matters where we return from the function because the compiler guarantees that when we do return, our code will always run the destructor for a, which releases the memory. Even if an exception is thrown, the destructor is guaranteed to run.

Scoping in C

In old versions of C, you could declare variables only at the top of a function or at the file level. You couldn't declare a variable within a `for` loop, for example:

```
void old_C_func() {
    int a;

    for (a = 0; a < 10; a++) {
        // OK
    }

    for (int b = 0; b < 10; b++) {
        // Not allowed! b is in block scope.
    }
}
```

Block scoping, as in this example, wasn't officially added to C until 1989 with the introduction of ANSI C, although some compilers may have supported it earlier. C has three main kinds of scope:

- *Function scope*—Variables declared at the function level
- *Block scope*—Variables declared within a code block
- *File scope*—Variables declared in a file

Variables within blocks can be nested and may be shadowed. The following code is valid:

```
void shadowing() {
    int a = 0;
    {
        int a = 1;
        printf("inner a=%d\n", a);
    }
    printf("outer a=%d\n", a);
}
```

> **(continued)**
>
> In this example, we shadow `a` by declaring it twice: once at the function level and again within a code block. If you were to run the code, it would print the following:
>
> ```
> inner a=1
> outer a=0
> ```
>
> Rust has block scoping, and you can shadow variables as well. Rust follows rules on scoping similar to those of modern C and C++, and it has some additional rules on handling moves, lifetimes, and borrowing. Notably, in Rust, a variable can outlive its declared scope if it's moved, which is a crucial difference from C and C++.

How does the compiler implement RAII? It does so through the use of the stack, which is scoped within a function or block, often denoted by curly braces ({ … }). Each new variable is pushed onto the stack when you enter a particular scope (such as a function). When you leave the scope, each variable is popped off the stack. The compiler has to store a little extra data alongside each variable so that it knows how to destroy each value safely. Still, the overhead is minimal and generally amounts to an additional pointer for anything that requires cleanup.

4.1.2 *A tour of RAII in Rust*

Object management in Rust follows the rules of RAII with two exceptions: unsafe code and `Copy` values. Variables must be initialized with a value at the time of declaration, and when a variable goes out of scope, it's destroyed with a destructor call (which we'll discuss later in this section).

Although the process of initializing variables and calling the destructor may be obscured by abstractions or layers of indirection, a variable must always be initialized with a value (unlike in C or C++, where you can have uninitialized variables), and an object's destructor is always called when it goes out of scope. For simple variables (such as those that aren't pointers, including `Rc` and `Arc`), Rust's borrow checker and move semantics make it relatively easy to reason about when variables or objects go out of scope and, thus, when they're destroyed.

Objects that are `Copy`—including primitive types such as integers, floats, and Booleans, and simple structures composed only of primitives—cannot have their destructors called because they're copied by value rather than being moved. The lack of destructors for `Copy` objects is a special case, and you'll need to be aware of it when you're working with `Copy` objects. You cannot define a destructor for a `Copy` object or rely on the destructor's being called for `Copy` objects. The trivial piece of code in the following listing illustrates the way RAII works.

Listing 4.1 Resource acquisition

```
fn main() {
    let status = String::from("Active");
```
← **String constructor acquires and allocates memory to hold a string**

```
        let statuses = vec![status];
        println!("{:?}", statuses);
    }
```

◁— Ownership of the string is passed into a Vec<String> upon initialization, and now the status string is held by statuses.

◁— statuses goes out of scope and releases both the Vec and String.

We can picture the construction part of this code, as shown in figure 4.1. Our new objects, which allocate memory on the heap behind the scenes (once for `String` and once for `Vec`), are created. The objects are initialized with the values we provide and then pushed onto the stack for the local scope. Because we transfer the ownership of the original `status` variable to `statuses`, the `status` on the stack effectively becomes an invalid reference. The Rust compiler handles this situation transparently, however, so we don't need to worry about it.

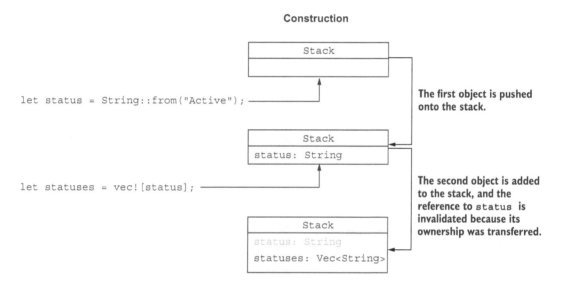

Figure 4.1   RAII entry and construction

We can picture the destruction part of this code, as shown in figure 4.2. Our new objects are destroyed one at a time as they're popped off the stack. For containers such as `Vec`, the destructor automatically calls the destructor for all children as well. So our original `status` string is destroyed along with `statuses`, although the original `status` reference is no longer valid because it has been moved.

In Rust, destruction is handled by an automatically generated destructor, which also recursively calls the destructor of every object member. The destructor first calls the `drop()` method for a given type, which is defined by the `Drop` trait as follows:

```
pub trait Drop {
    fn drop(&mut self);
}
```

Destruction

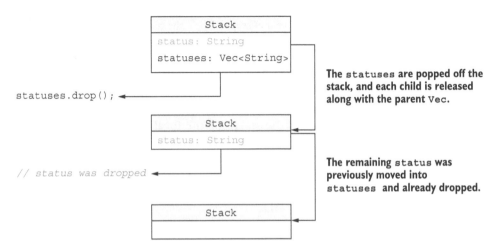

Figure 4.2   RAII exit and destruction

If you implement `Drop` for any type, its corresponding `drop()` method is guaranteed to be called whenever a variable of that type goes out of scope. Then the automatic destructor recursively calls the destructors of every member variable.

Rust always calls the destructors for all objects whenever they go out of scope, so you don't need to call `drop()` manually. Also, you can't override this behavior without using `unsafe`. (That is, you can't stop Rust from calling destructors.)

4.1.3   *Summarizing RAII in Rust*

There are a few key points to remember about RAII in Rust, much of which will be intuitive to anyone who's familiar with RAII (as in C++ and other languages):

- *RAII is used extensively in Rust.*
  - Rust does not feature garbage collection; memory management is explicit. Allocating memory on the heap is normally accomplished with `Box` or `Vec`.
  - Object lifetimes are deterministic and known at compile time (except when you're using smart pointers).
  - Stack-allocated objects follow the same RAII rules as heap-allocated ones.
- *Memory management objects use RAII.*
  - `Box` and `Vec` use RAII to acquire, initialize, and release memory resources.
  - Smart pointers such as `Rc` and `Arc` use RAII to implement reference counting each time a pointer is cloned and destroyed.
  - `RefCell` returns the borrow references `Ref` and `RefMut`, which use RAII to guard against multiple simultaneous references.

- *Several synchronization primitives use RAII.*
  - `Mutex::lock()` returns a `MutexGuard` on success. `MutexGuard` is an RAII-based lock guard that automatically unlocks the mutex when it's destroyed.
  - `RwLock` returns `RwLockReadGuard` or `RwLockWriteGuard` when you acquire shared read or exclusive write access, respectively, for a read-write lock.
  - `Condvar` requires a `MutexGuard` to wait on a condition variable, as shown in listing 4.2.

To demonstrate RAII in Rust, we'll use `Mutex` and `Condvar` to create a simple threaded example, which involves creating one thread that increments a value and notifies the main thread when it's done.

> **Listing 4.2   RAII in Rust with `Mutex` and `Condvar`**

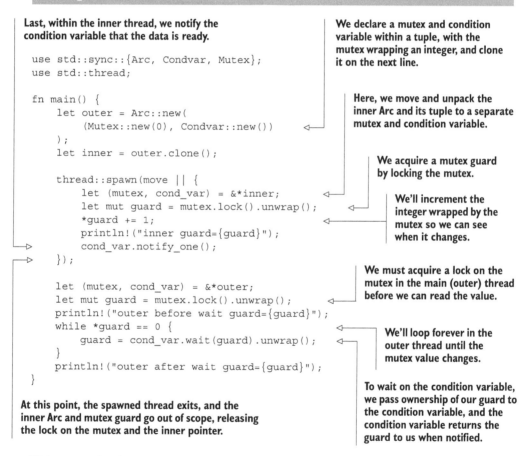

Last, within the inner thread, we notify the condition variable that the data is ready.

We declare a mutex and condition variable within a tuple, with the mutex wrapping an integer, and clone it on the next line.

```
use std::sync::{Arc, Condvar, Mutex};
use std::thread;

fn main() {
    let outer = Arc::new(
        (Mutex::new(0), Condvar::new())
    );
    let inner = outer.clone();

    thread::spawn(move || {
        let (mutex, cond_var) = &*inner;
        let mut guard = mutex.lock().unwrap();
        *guard += 1;
        println!("inner guard={guard}");
        cond_var.notify_one();
    });

    let (mutex, cond_var) = &*outer;
    let mut guard = mutex.lock().unwrap();
    println!("outer before wait guard={guard}");
    while *guard == 0 {
        guard = cond_var.wait(guard).unwrap();
    }
    println!("outer after wait guard={guard}");
}
```

Here, we move and unpack the inner Arc and its tuple to a separate mutex and condition variable.

We acquire a mutex guard by locking the mutex.

We'll increment the integer wrapped by the mutex so we can see when it changes.

We must acquire a lock on the mutex in the main (outer) thread before we can read the value.

We'll loop forever in the outer thread until the mutex value changes.

To wait on the condition variable, we pass ownership of our guard to the condition variable, and the condition variable returns the guard to us when notified.

At this point, the spawned thread exits, and the inner Arc and mutex guard go out of scope, releasing the lock on the mutex and the inner pointer.

This example demonstrates multiple simultaneous uses of RAII—enough to make anyone's head spin. To summarize:

- `Mutex` wraps an arbitrary value (in this case, an integer, but we could wrap any object in a mutex) that is released when it goes out of scope

- `Mutex` and `Condvar` use the `MutexGuard`'s RAII to hand off a locked mutex.
- `Arc` provides a thread-safe reference-counted pointer to our mutex and condition variable.

When the inner thread exits, the `MutexGuard` is released, which unlocks the mutex, and the `Arc` is dropped, which releases the pointer to the mutex and condition variable. The outer thread simultaneously acquires the lock on the mutex, waits for the condition variable to be notified, and releases the lock when the guard goes out of scope. Note that we don't know which thread will run first, so we must wait for the condition variable to be notified before we can proceed, and we can't guarantee the order of execution.

RAII is a powerful pattern that allows us to manage resources safely and handle cleanup automatically. Rust's strict rules on ownership and borrowing make it easy to reason about when objects go out of scope and when their destructors will be called.

4.2   *Passing arguments by value vs. reference*

At first glance, this topic may appear to be basic or entry-level. After spending some time writing Rust code, however, you'll realize that it's imperative to think carefully about whether you want to pass arguments by value or reference. A lot of nuance is involved, but I'll provide some guidance on how to use the common patterns and when to do what.

4.2.1   *Passing by value*

In Rust, passing arguments by value typically constitutes a move. In simple terms, a *move* occurs when you transfer the ownership of an object from one scope to another. A move could occur when a function is called, a closure is created, an object is assigned, or a value is returned from a function. Another interesting property of passing by value is the fact that it respects RAII. The simple code sample in the following listing illustrates passing by value.

Listing 4.3   Reversing a string, passing by value

```
fn reverse(s: String) -> String {
    let mut v = Vec::from_iter(s.chars());       ← Constructs a Vec from an iterator
    v.reverse();                                     over the characters in s
    String::from_iter(v.iter())          ←           Reverses the newly
}                                                    constructed vector in place
       Returns a new string from an iterator over
       the reversed characters in our vector v
```

This code is an example of a function that reverses the characters in a string. The function takes a string by value and returns a new string. We can test our function as follows, ensuring that the returned value is the reverse of the one provided:

```
assert_eq!("abcdefg", reverse(String::from("gfedcba")));
```

Sometimes, it's handy to move values into a function and immediately move them back out, as in the preceding example. We might do this to avoid borrowing or cloning a value. If we have multiple values to return, we can return a tuple instead:

```
fn reverse_and_uppercase(s: String) -> (String, String) {
    let mut v = Vec::from_iter(s.chars());
    v.reverse();
    let reversed = String::from_iter(v.iter());
    let uppercased = reversed.to_uppercase();
    (reversed, uppercased)
}
```

Although this example has only one argument passed into the function, we could easily pass in multiple arguments by value and return multiple values. We can test the code as follows:

```
assert_eq!(
    reverse_and_uppercase("abcdefg".to_string()),
    ("gfedcba".to_string(), "GFEDCBA".to_string())
);
```

Don't be afraid to pass by value, but be aware that it performs a move for any type that doesn't implement `Copy`, although this can be advantageous sometimes.

4.2.2 Passing by reference

You obtain a reference to an object or variable by borrowing it. You can think of it as behaving similarly to a pointer except that you can't perform bitwise or arithmetic operations on a reference, and you can pass a reference or assign it only once without manipulating the reference after assignment. References are denoted by a `&` prefixing the type specifier and may include lifetimes, represented by a single quote (`'`) following the `&` and an optional lifetime identifier such as `&'a String`. References can be immutable (`&str`, the default) or mutable (`&mut String`). Let's rewrite our reverse function, but this time, we'll pass the input by reference, as shown the following listing.

Listing 4.4   Reversing a string, passing by reference

```
fn reverse(s: &str) -> String {
    let mut v = Vec::from_iter(s.chars());
    v.reverse();
    String::from_iter(v.iter())
}
```

You may notice immediately that these two functions have only one difference: the argument `s: String` has been swapped for `s: &str`. When we test our code, however, we can do things slightly differently:

```
assert_eq!("abcdefg", reverse("gfedcba"));
```

Notice that instead of creating a string with `String::from()`, we can pass a static string (such as `&'static str`). This approach is nice and a little more ergonomic. If we wanted to do so, we could call our reverse function as follows:

```
assert_eq!(
  "race car", reverse(&String::from("rac ecar"))
);
```

◁— **Borrowing a String gives us &str because String implements the Borrow and BorrowMut traits to return &str and &mut str, respectively.**

What if we want to update our string in place? This process is a little trickier because we can't easily perform a proper (zero-copy) in-place reversal. We can emulate the behavior as shown in the following listing. This code provides suitable performance at the expense of some temporary memory overhead.

Listing 4.5   Reversing a string in place (sort of)

```
fn reverse_inplace(s: &mut String) {
    let mut v = Vec::from_iter(s.chars());
    v.reverse();
    s.clear();
    v.into_iter().for_each(|c| s.push(c));
}
```

We can test our in-place reversal like this:

```
let mut abcdefg = String::from("gfedcba");
reverse_inplace(&mut abcdefg);
assert_eq!("abcdefg", abcdefg);
```

Why is it impossible to mutate Rust strings in place?

You may have noticed that in-place string manipulation in Rust isn't easy. The reason is simple: strings in Rust are always valid UTF-8, which means that characters could span multiple bytes or be composed of grapheme clusters in the Unicode standard.

A *grapheme* is the smallest unit of a writing system, which could be an ordinary character (such as the letter a), or a character that includes an accent such as é or an emoji character. When we think about strings and characters, we tend to believe that one displayed character equals 1 byte, which is true only of strict ASCII characters.

Because grapheme clusters can span multiple Unicode characters and multiple bytes, it's quite complicated to handle them correctly, so the Rust standard library does not support handling them directly. Instead, you need to use a crate such as `unicode-segmentation` (https://crates.io/crates/unicode-segmentation).

If you need to update a string in place by manipulating its bytes, you have two options:

- You can use the `std::mem::take` function to gain access to the underlying bytes of a string and manipulate a buffer directly.

- You can use an unsafe method, such as `String::as_mut_vec()` or `str::as_bytes_mut()`, which returns references to the underlying bytes.

The first method is preferred, as it doesn't require unsafe code, but in either case, you need to consider how to handle UTF-8 characters safely. If you try to manipulate the bytes of a string directly, you may get some peculiar results.

4.2.3 When to do what: Passing by value vs. reference

It may not be clear at first when to pass a value by reference or by value by using a move, so I'll provide some general guidelines to help you develop some intuition. As with anything, practice will help you get a sense of which pattern is correct in which situations. If you consider yourself to be intermediate or advanced in Rust (or a language with similar semantics), this fact may be obvious. Still, it may be beneficial to formalize these ideas and perhaps make some new neural connections, which could be valuable in light of the current trendiness of neural networks.

To add one more layer of complexity, keep in mind that object methods typically take `self` as an argument, still following the same rules: `self` can be passed by value (performing a move) or by reference (no move), which, as we'll discuss throughout this book, can create some interesting patterns that are somewhat unique to Rust. To begin, let's look at different ways to handle arguments (table 4.1).

Table 4.1 Summary of argument passing

Argument passed by	Prefix	Moved?	Ownership	Default use case
Reference	`&`	No	Retained by caller	The callee requires temporary access to a value.
Mutable reference	`&mut`	No	Retained by caller	The callee needs to mutate a value without ownership.
Value	N/A	Yes	Transferred to callee	The callee needs to obtain ownership of the value.
Mutable value	`mut`	Yes	Transferred to callee	The callee needs to obtain ownership and mutate the value.

Most of the time, you want to pass arguments by reference. Generally, you wouldn't want to pass by reference in only two cases: when you use primitive types (such as `i32` and `usize`) and when you need to move the ownership of a value into the callee and possibly return the same value as part of a new object or on its own. But you need to think carefully about why you're transferring ownership. Do you want to mutate the value? If so, is there a reason why you can't use a reference (such as to avoid copies or method chaining)?

To help you evaluate what to do, I created the simple flowchart shown in figure 4.3. You can refer to this flowchart for guidance on handling argument passing in most cases, at least until the process becomes second nature.

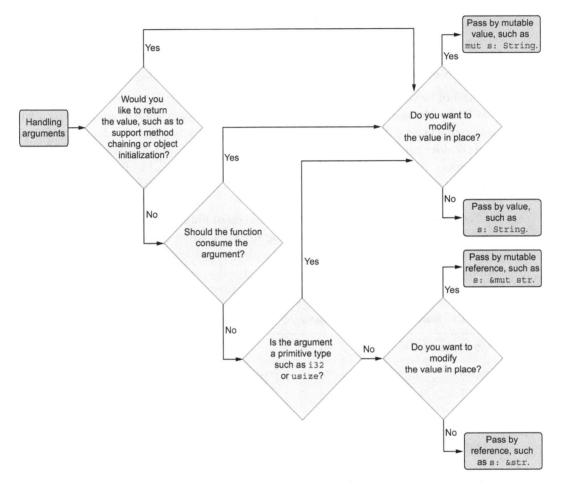

Figure 4.3   Deciding how to handle arguments

4.3   *Constructors*

Strictly speaking, Rust does not have a formal notion of a constructor in the same way that languages such as C++, C#, and Java do. In Rust, a *constructor* is merely a design pattern in which you create a method, typically called new(), that accepts any number of initialization arguments and returns a new object immediately after creation. Although Rust has no new keyword, it can help you to think about it as being equivalent to new in other languages. But understand that (as already mentioned) Rust has no formal concept of a constructor; therefore, any constructors you find in Rust are strictly conventional patterns, not special methods, as they are in C++, Java, and C#.

The following listing illustrates a simple constructor by creating a container to model a pizza with toppings. Note that the constructor doesn't provide a way to add toppings (yet).

Listing 4.6 Modeling a pizza pie

```
#[derive(Debug, Clone)]
pub struct Pizza {
    toppings: Vec<String>,
}

impl Pizza {
    pub fn new() -> Self {               <──  This constructor
        Self { toppings: vec![] }             takes no arguments
    }                                          and returns Self
}                                              (an empty Pizza).
```

We can make an empty `Pizza`:

```
let pizza = Pizza::new();
println!("pizza={:?}", pizza);
```

Running the preceding code produces the following output:

```
pizza=Pizza { toppings: [] }
```

For simple cases, you'll likely want to initialize objects with some values, perhaps derived from constructor arguments. In Rust, `new()` typically takes no arguments and returns an empty object, as is the case with `Vec::new()`, which returns an empty vector.

There's no rule against including initialization arguments with `new()`, but it's common to implement the `From` trait instead when you want to create a new object from another object. This approach makes sense only when a 1:1 mapping exists (when `String::from(…)` constructs a new string, for example). Let's rewrite our constructor from listing 4.6 so that we can initialize our pizza's toppings.

Listing 4.7 A better pizza constructor

```
impl Pizza {
    pub fn new(toppings: Vec<String>) -> Self {    <──  Our constructor takes a
        Self { toppings }             <──                Vec with our toppings
    }                                                    and moves those
}            Because the name of the constructor         toppings into our newly
             argument and the Pizza member are the same,  constructed pizza.
             we can shorten toppings: toppings to toppings.
```

Let's test our new constructor:

```
let pizza = Pizza::new(vec![
    String::from("tomato sauce"),
```

```
        String::from("mushrooms"),
        String::from("mozzarella"),
        String::from("pepperoni"),
    ]);
    println!("pizza={:#?}", pizza);
```

When we test our pizza, which is likely to taste a lot better, the code prints the following:

```
pizza=Pizza {
    toppings: [
        "tomato sauce",
        "mushrooms",
        "mozzarella",
        "pepperoni",
    ],
}
```

Because Rust doesn't permit function overloading, you can create only one method called `new()`, so think carefully about what you want this function to do. In most cases, the function should provide the minimally necessary behavior, such as returning a new empty object (as with `Vec::new()`) with the minimum required arguments. Also, some people create additional constructors that begin with `new_` and take additional arguments. `Vec::new_in(alloc: A)` (available in nightly Rust only), for example, accepts an optional memory allocator and returns an empty `Vec` that uses the specified allocator.

> **NOTE** If your set of initialization arguments grows in complexity, you probably want to use the builder pattern, which we'll discuss in chapter 5.

4.4   *Object-member visibility and access*

Rust generally defaults to private visibility. Optionally, you can make entities public with the `pub` keyword. Public visibility has slightly different meanings depending on the context. Here, we'll discuss object members, in which case adding the `pub` keyword means that they can be accessed or modified directly. Let's revisit the pizza example from section 4.3.1, this time making the toppings public outside the current module.

Listing 4.8 A pizza with `pub toppings`

```
#[derive(Debug, Clone)]
pub struct Pizza {
    pub toppings: Vec<String>,        ◁— Note the pub
}                                        visibility specifier.
```

In effect, this code allows us to treat the `toppings` member as a plain variable and do things like this:

```
let mut pub_pizza = Pizza {
    toppings: vec![String::from("sauce"), String::from("cheese")],
};
```

```
// Remove the last topping.
pub_pizza.toppings.remove(1);
println!("pub_pizza={:?}", pub_pizza);
```

If we run this code, we'll get the following output:

```
pub_pizza=Pizza { toppings: ["sauce"] }
```

When would you want to do this? Generally, you *wouldn't* want to do this except when you have data containers with no methods and their only purpose is to contain data. Most of the time, you want to control access to members with *accessors* (methods that fetch private members) and modify members with *mutators* (methods that allow you to mutate private members). Accessors and mutators are often called *getters* and *setters*, though in Rust, it's important to distinguish between setting a value (such as a move) and mutating a value in place.

> **TIP**   A bit of boilerplate is involved in these methods, but tools like `rust-analyzer` make generating getters and setters for each member easy. My book *Code Like a Pro in Rust* (Manning, 2024; https://www.manning.com/books/code-like-a-pro-in-rust) has a section on `rust-analyzer`, but you can refer to the generators documentation at https://mng.bz/5lV8 for details.

Let's update our pizza by changing the `toppings` back to private (we don't want them to be public) and adding an accessor, a mutator, and a setter.

Listing 4.9   Providing access to our pizza `toppings`

```
impl Pizza {
    pub fn toppings(&self) -> &[String] {        ⟵  Our accessor or getter returns
        self.toppings.as_ref()                       a slice of the toppings vector.
    }

    pub fn toppings_mut(&mut self) -> &mut Vec<String> {   ⟵  Our mutator returns
        &mut self.toppings                                     a mutable reference
    }                                                          to the underlying
                                                               toppings vector.
    pub fn set_toppings(&mut self, toppings: Vec<String>) {   ⟵
        self.toppings = toppings;
    }                                      Our setter takes a new vector by
}                                          value and replaces the existing
                                           one with the new one.
```

In this example, each method takes a reference to `self`, and the mutable methods take a mutable reference. Note that I'm returning a slice from the underlying `Vec` instead of a direct reference. Returning the data as a `Vec` and returning it as a slice are equivalent, but merely returning a slice is slightly more idiomatic because a slice is generally used to represent immutable contiguous sequences (such as the lowest common denominator).

We could also modify `set_toppings()` slightly so that it returns (or moves) the existing `toppings` while replacing the current ones. We may want to call that modification something like `replace_toppings()`.

> **Listing 4.10   Providing a method to swap the `toppings`**

```
impl Pizza {
    pub fn replace_toppings(
        &mut self,
        toppings: Vec<String>,
    ) -> Vec<String> {
        std::mem::replace(&mut self.toppings, toppings)
    }
}
```

This code uses `std::mem::replace()`, allowing us to swap by replacing the existing `toppings` with a move and returning the old `toppings` with another move. This approach prevents cloning and duplication, which is an excellent little optimization.

4.5   *Error handling*

Handling errors in Rust is surprisingly uncomplicated. Common practice is to lean heavily on Rust's `Result`, which has special support in the language for the `?` operator, as demonstrated later in this section.

Before I get into code samples, I should discuss the two sides of error handling: producing errors (such as a function that might return an error) and handling results (what to do when a function returns an error).

When it comes to producing errors, we typically use plain structs or enums that contain the necessary error metadata (error type, messages, and so on). The standard library provides a few error types (such as `std::io::Error`) that you can use, but often, you merely include them within your own error types (as enum variants, for example) or return them directly unchanged. Creating your own error type is as simple as defining any struct or enum; then you can return that error within a `Result`. The standard library also has an error trait, `std::error::Error`, that you can implement for your own error types, but its use is optional. In practice, implementing `std::error::Error` for custom error types is uncommon.

Handling errors typically involves using a combination of two strategies: explicitly handling each case with pattern matching (or some other control flow) or letting the errors bubble up to the caller. For the latter option, we can sometimes get away with using the `?` operator. Using the `?` operator is simple: postfix any function call that returns `Result` or `Option`, and the `?` operator unwraps the result for you while returning from your function early if there's an error or `None` (in the case of `Option`). This approach can be convenient because it lets us chain function calls that may return errors (or `None` in the case of `Option`). The downside to using `?` is that it can be lazy. Sometimes, you should handle errors and take action explicitly. When you use `?`, you'll likely need to implement the `From` trait for your error types, creating another place where you can mix in your error-handling logic.

Let's write a function that reads the *nth* line from a file and returns that line as a string. As we'll see, this function can fail in several ways, so we'll need to handle each case. The following listing shows our first attempt.

Listing 4.11   Reading the *nth* line from a file

Our error type is an enum
with two possible values.

Error::Io contains
std::io::Error, which we
bubble up to the caller.

```
use std::path::Path;

#[derive(Debug)]
pub enum Error {
    Io(std::io::Error),
    BadLineArgument(usize),
}
```

Error::BadLineArgument is the
error we'll return when the line
number is invalid.

```
impl From<std::io::Error> for Error {
    fn from(error: std::io::Error) -> Self {
        Self::Io(error)
    }
}
```

We implement From to allow
conversion of std::io::Error
to our error type Error.

```
fn read_nth_line(path: &Path, n: usize) -> Result<String, Error> {
    use std::fs::File;
    use std::io::{BufRead, BufReader};
    let file = File::open(path)?;

    let mut reader_lines = BufReader::new(file).lines();
    reader_lines
        .nth(n - 1)

        .map(|result| result.map_err(|err| err.into()))
        .unwrap_or_else(|| Err(Error::BadLineArgument(n)))
}
```

Here, we use the ?
operator to obtain
a file handle.

Last, if the value returned is None, we
read past the end of the file before
hitting the target number of lines.

Note that we must subtract one
from n here because the first line
is the 0th line read from the file.

The nth() method returns
Option<Result<String,
std::io::Error>>, so we need to convert
the contained error to our error type.

BufReader gives us a buffered reader for our file
handle, and the BufRead trait provides lines(),
which gives us an iterator over each line in the file.

Our function `read_nth_line()` uses the `std::io::BufRead` trait, which gives us several handy features, including the `lines()` method, which returns an iterator over each line in the file. Let's test our function with the following code:

```
let path = Path::new("Cargo.toml");
println!(
    "The 4th line from Cargo.toml reads: {}",
    read_nth_line(path, 4)?
);
```

Running this code produces the following output:

```
The 4th line from Cargo.toml reads: edition = "2021"
```

A subtle bug is introduced on the line where we subtract one from n when calling `nth()`. An overflow will occur if n is 0, so we need to handle that bug. Note that if the program is compiled in release mode, the overflow will be suppressed, as Rust checks for integer overflows only when code is compiled in debug mode; otherwise, it mimics the behavior of C.

We have a few options for handling this case, but we'll do a check on n and return early with an error if the value is less than 1.

Listing 4.12 **Reading the *nth* line from a file**

```
fn read_nth_line(path: &Path, n: usize) -> Result<String, Error> {
    if n < 1 {
        return Err(Error::BadLineArgument(0));
    }
    use std::fs::File;
    use std::io::{BufRead, BufReader};
    let file = File::open(path)?;

    let mut reader_lines = BufReader::new(file).lines();
    reader_lines
        .nth(n - 1)
        .map(|result| result.map_err(|err| err.into()))
        .unwrap_or_else(|| Err(Error::BadLineArgument(n)))
}
```

Next, we should write some unit tests for our function to verify that it behaves as expected. We'll write the tests as shown in the following listing.

Listing 4.13 **Unit tests for reading the *nth* line from a file**

```
#[cfg(test)]
mod tests {
    use super::*;
    #[test]
    fn test_can_read_cargotoml() {
        let third_line = read_nth_line(Path::new("Cargo.toml"), 3)
            .expect("unable to read third line from Cargo.toml");
        assert_eq!("version = \"0.1.0\"", third_line);
    }
    #[test]
    fn test_not_a_file() {
        let err = read_nth_line(Path::new("not-a-file"), 1)
            .expect_err("file should not exist");
        assert!(matches!(err, Error::Io(_)));
    }
}
```

Tries to read from the file not-a-file, which doesn't exist and returns an I/O error

We check the error returned using matches!, which allows us to provide a pattern to match against and returns a Boolean we can assert on.

```
        #[test]
        fn test_bad_arg_0() {
            let err = read_nth_line(Path::new("Cargo.toml"), 0)
                .expect_err("0th line is invalid");
            assert!(matches!(err, Error::BadLineArgument(0)));
        }
        #[test]
        fn test_bad_arg_too_large() {
            let err = read_nth_line(Path::new("Cargo.toml"), 500)
                .expect_err("500th line is invalid");
            assert!(matches!(err, Error::BadLineArgument(500)));
        }
    }
```

Again, we check that the error returned matches the expected value.

Here, we check the error returned using a pattern match with a specific value (0).

This project's Cargo.toml has only 8 lines, so 500 is well beyond the end of the file and should result in an error.

Here, we check whether passing a zero-value for n produces an error.

As we've seen, dealing with errors is not complicated in Rust. In most cases, you'll want to create an error type for your library or application to encapsulate all the errors it may return, and in many cases, you simply want to return the underlying error unaltered.

4.6   Global state

There comes a time in every developer's life when they need to deal with global state. We tend to avoid global state for good reason; it can introduce race conditions, corruption risk, poor separation of concerns, and a host of other problems. As hard as you try to avoid doing so, however, you'll eventually need to deal with a situation in which you require global state. In this section, I discuss some strategies for handling global state in Rust, which involves challenges and advantages owing to Rust's memory and ownership models.

Global state is sometimes implemented via the singleton pattern, which some developers consider to be an antipattern. We'll discuss this topic again in chapter 10, but I'll say here that you should use global state (and singletons) sparingly.

Now let's talk about global variables in Rust. Rust allows only two kinds of global variables: `static` and `const`. In both cases, the variable's value must be determined at compile time. In other words, you can't perform run-time initialization with any global variables. You can define mutable static variables, which allow us to modify their values at run time, but this approach is considered to be unsafe and requires the use of the `unsafe` keyword. Static variables must also be `Sync`—in other words, to allow thread-safe access (to prevent race conditions). Additionally, allocations are not permitted in statics (you can't use anything that allocates memory on the heap), and the `drop()` method from `Drop` is never called at shutdown when you use static variables.

Because of these limitations, it's common to use some form of lazy just-in-time initialization for global state. Several crates make this task easy, but before we examine those crates, let's discuss how we could perform it manually.

We can't create a static vector of strings because both `Vec` and `String` are heap-allocated. The following code won't compile:

```
static POPULAR_BABY_NAMES_2021: Vec<String> = vec![
    String::from("Olivia"),
    String::from("Liam"),
    String::from("Emma"),
    String::from("Noah"),
];
```

Trying to compile this code will produce a long list of errors:

```
error[E0010]: allocations are not allowed in statics
  --> src/main.rs:1:47
   |
1  |     static POPULAR_BABY_NAMES_2021: Vec<String> = vec![
   |  _____^
2  | |       String::from("Olivia"),
3  | |       String::from("Liam"),
4  | |       String::from("Emma"),
5  | |       String::from("Noah"),
6  | | ];
   | |_^ allocation not allowed in statics
   |
   = note: this error originates in the macro `vec` (in Nightly builds,
 run with -Z macro-backtrace for more info)

error[E0015]: cannot call non-const fn `<String as From<&str>>::from`
 in statics
  --> src/main.rs:2:5
   |
2  |       String::from("Olivia"),
   |       ^^^^^^^^^^^^^^^^^^^^^^^
   |
   = note: calls in statics are limited to constant functions, tuple
 structs and tuple variants
   = note: consider wrapping this expression in `Lazy::new(|| ...)`
 from the `once_cell` crate: https://crates.io/crates/once_cell
```

You may have noticed the suggestion that we use `once_cell` from the compiler error, which we'll do in a moment. First, let's see whether we can make this approach work without using crates.

To create a static global variable, we need to use the `std::thread_local!` macro, which provides thread-local storage that's `Sync` (thread-safe). Thread-local storage enables us to store data that is local to the current thread but also makes memory globally accessible.

We need to use a reference counted pointer, `Arc`, and `Mutex` to share the inner data safely. Last, our `Vec<String>` must be wrapped in an `Option` because we can't initialize a `Vec` or `String` at compile time. In this case, the pointer we use to access the data is local to the current thread, but the data itself is global, so we're left with the code in the following listing.

Listing 4.14   Declaring a thread-local, global-static variable

```
thread_local! {
    static POPULAR_BABY_NAMES_2021: Arc<Mutex<Option<Vec<String>>>> =
        Arc::new(Mutex::new(None));
}
```

We need to initialize our `Vec` with some data. Somewhere in our code, such as `main()`, we have to do the following to initialize the data.

Listing 4.15   Initializing a thread-local, global-static variable

```
let arc = POPULAR_BABY_NAMES_2021.with(|arc| arc.clone());
let mut inner = arc.lock().expect("unable to lock mutex");
*inner = Some(vec![
    String::from("Olivia"),
    String::from("Liam"),
    String::from("Emma"),
    String::from("Noah"),
]);
```

This approach is a rather unpleasant way to handle this situation. Also, we have to be extra careful to initialize our global data correctly, in the proper order, before anything else might try accessing values.

In practice, you shouldn't handle global state this way. Instead, I suggest using one of several crates that provide this behavior in a nice API (table 4.2).

Table 4.2   Summary of global state crates

Crate	Repository	Downloads as of March 3, 2024	Description
lazy-static.rs	https://mng.bz/oegy	215,759,981	Macro for declaring lazily evaluated statics
once_cell	https://github.com/matklad/once_cell	213,996,727	Provides two new cell-like types that can be used to initialize global state
static_init	https://gitlab.com/okannen/static_init	3,391,550	Provides global statics with higher performance and several features including dropping data

In the following sections, we'll implement the example shown in listings 4.14 and 4.15, using the crates from table 4.2.

4.6.1   *lazy-static.rs*

The `lazy-static.rs` crate is the most popular way to solve the global state problem in Rust (as of this writing). Its API is based on a simple macro that uses the `static ref` syntax to define global variables, with an option to use a closure to perform initialization. Using this crate, we can initialize some global state.

Listing 4.16 Popular baby names with `lazy-static.rs`

```rust
use lazy_static::lazy_static;

lazy_static! {
    static ref POPULAR_BABY_NAMES_2020: Vec<String> = {
        vec![
            String::from("Olivia"),
            String::from("Liam"),
            String::from("Emma"),
            String::from("Noah"),
        ]
    };
}
```

If you want the data to be mutable, you could use `Mutex<Vec<String>>` or `RwLock<Vec<String>>`, but for this example, we'll treat this data as immutable. We can test our code with the following:

```rust
println!("popular baby names of 2020: {:?}", *POPULAR_BABY_NAMES_2020);
```

Note that we only have to dereference the variable using the * operator to access its value because `lazy-static.rs` provides the `Deref` trait. Running the preceding code results in the following output:

```
popular baby names of 2020: ["Olivia", "Liam", "Emma", "Noah"]
```

4.6.2 *once_cell*

The `once_cell` crate is rapidly gaining in popularity; it provides a more generic API for handling global state than `lazy-static.rs`. For this reason, I recommend using `once_cell` instead of `lazy-static.rs` for new projects. But if you're already using `lazy-static.rs` or are more familiar with it, it's an excellent solution.

Let's implement the same thing with `once_cell`. The following listing has a nice, concise API.

Listing 4.17 Popular baby names with `once_cell`

```rust
use once_cell::sync::Lazy;

static POPULAR_BABY_NAMES_2019: Lazy<Vec<String>> = Lazy::new(|| {
    vec![
        String::from("Olivia"),
        String::from("Liam"),
        String::from("Emma"),
        String::from("Noah"),
    ]
});
```

The `once_cell::sync::Lazy` API provides the `Deref` trait so that we can access the values with the * operator:

```
println!("popular baby names of 2019: {:?}", *POPULAR_BABY_NAMES_2019);
```

As with `lazy-static.rs`, we can wrap the data with `Mutex` or `RwLock` to enable mutability.

4.6.3   *static_init*

Last, we'll look at `static_init`, which has a few nice features and excellent performance.

Listing 4.18   Popular baby names with `static_init`

```
use static_init::dynamic;

#[dynamic]
static POPULAR_BABY_NAMES_2018: Vec<String> = vec![
    String::from("Emma"),
    String::from("Liam"),
    String::from("Olivia"),
    String::from("Noah"),
];
```

To enable mutability, we can add the `mut` keyword (such as `static mut POPULAR_BABY_NAMES_2018 …`). `static_init` also provides `Deref`, like `lazy-static.rs` and `once_cell`, so we can access the value like so:

```
println!("popular baby names of 2018: {:?}", *POPULAR_BABY_NAMES_2018);
```

4.6.4   *std::cell::OnceCell*

I should note that the Rust standard library (as of Rust 1.70) includes `std::cell::OnceCell` and `std::sync::OnceLock`, which partially solve the static initialization problem but without convenient lazy initialization at the global level. An experimental API called `std::cell::LazyCell` is available for this feature, but it's not yet available in stable Rust. Using `std::cell::OnceCell` is roughly equivalent to using the `thread_local!` macro, as discussed earlier in this chapter.

You can create a global instance of `std::cell::OnceCell`, but you can't initialize its value at the global scope within a single expression. If you prefer to avoid using crates for this task, this may be an acceptable tradeoff. The main downside to separating the declaration and initialization is that this approach decreases clarity and could result in duplicate initialization code or a potential race condition if multiple paths to competing initialization code exist.

For the sake of completeness, we can implement the equivalent behavior by using `std::cell::OnceCell`, but our initialization must occur within a function as opposed to the global context. The following listing simply places the code within `main()`.

Listing 4.19   Using `std::cell::OnceCell`

```
let popular_baby_names_2017: OnceCell<Vec<String>> = OnceCell::new();
popular_baby_names_2017.get_or_init(|| {
    vec![
        String::from("Emma"),
        String::from("Liam"),
        String::from("Olivia"),
        String::from("Noah"),
    ]
});
```

Summary

- RAII is used extensively in Rust, and it works well in conjunction with Rust's move semantics to safely handle ownership, resource release, and synchronization.

- You can use RAII to build data structures and containers that manage resources safely and perform cleanup by implementing the `Drop` trait.

- Function-call arguments can be passed by value or reference. Passing by value in particular enables some unique patterns in Rust.

- Arguments passed by value are moved from the caller's context into the callee's context and can be returned from the callee to the caller.

- Object members are private by default. We commonly write methods to access or mutate member values as opposed to using public members, except when structures are used strictly as data containers and direct access is preferred.

- Functions that might produce errors should return `Result`, and we generally create error types to contain the details on any errors we might return.

- We can use the `?` operator to keep code tidy without handling every error case explicitly.

- By implementing the `From` trait for our error types, we can handle a variety of error cases gracefully.

- Handling global state in Rust is tricky, but several crates make the task easy. The `once_cell` crate, for example, provides a concise API for lazy initialization and global state.

Design patterns: Beyond the basics

This chapter covers

- Metaprogramming with macros
- Implementing the builder pattern in Rust
- Building fluent interfaces
- Observing the observer pattern
- Understanding the command pattern
- Exploring the newtype pattern

Chapters 2 and 3 introduced the core Rust building blocks: generics, traits, pattern matching, and functional programming features. In this chapter, we're going to build on what we learned in those chapters by exploring those themes further and examining design patterns in Rust.

Using what we've learned, we can start to build more concrete patterns in a way that is consistent with Rust idioms. Although we won't explore all the possible patterns, I'll present carefully chosen examples that demonstrate the fundamentals needed to build nearly any design pattern in Rust.

If generics, traits, pattern matching, and closures are the raw ingredients of any design pattern, the patterns in this chapter represent archetypes of nearly any other pattern that combines those features. Before diving right into the patterns themselves, we'll discuss macros, which aren't patterns themselves but are often used in advanced design patterns.

Although it may seem odd to introduce macros at this point, it's critical to understand macros in Rust before proceeding; they're widely used, and you won't get too far without having some basic understanding of them. Also, we'll use macros in chapter 8 to see how these features interact, compound, and provide code support.

Four of the five patterns discussed in this chapter are commonly found across languages, libraries, and SDKs: builders, fluent interfaces, observers, and the command pattern. The last pattern we'll look at is newtype, which is Rust-specific. These patterns are popular for good reason: they provide well-understood, useful, widely applicable abstractions for common programming challenges. Even if you never implement these patterns, after you recognize them, you'll see them everywhere.

5.1 Metaprogramming with macros

Macros are tools for metaprogramming, typically using a preprocessor. Macros let you extend or augment the features of a programming language. *Metaprogramming* is the process of using code to generate code, and *preprocessing* is the process of executing code (or macros) *before* the code is compiled.

Macros are often provided as a domain-specific language (DSL) for generating code before the compiler runs. Macros can also be bolted onto a language by means of a custom preprocessing step, which we sometimes see in languages that don't support macros.

Many languages feature macros, including C and C++, which have a basic but useful macro system. Lisp is well known for its advanced macro system. Elixir, Erlang, Scala, and OCaml also have macro systems.

Compared with those in C and C++, Rust's macros are quite a bit more advanced (and more complex, for that matter). The macros in C and C++ rely on textual substitution and provide little in terms of type checking, parameter matching, or even proper scoping.

Rust's macros are special in that they're type-safe, which makes them easier to work with and safer to use than those in other systems. Rust's compiler also does a good job of providing helpful error messages with macros, although things can get hairy if the macros become too complex. In addition to checking types, Rust's macros check for hygiene to ensure that the variables and identifiers used in the macro don't conflict with the variables and identifiers in the calling code.

Macros can be great complements to any codebase, especially for code that is verbose and repetitive. Like any other tools, macros can be misused, and they may also be used to mask code smells.

One more thing before we move on—Rust currently offers two macro systems: the regular macro system, *declarative macros*, available in Rust by default, and *procedural macros*, which require a flag to enable. Procedural macros are much more complex but offer a great deal more in terms of features and flexibility. We'll revisit procedural macros in chapter 8. This section discusses declarative macros.

5.1.1 A basic declarative macro in Rust

Let's take a look at a basic macro. In Rust, you can recognize a macro by the ! symbol after a keyword. Calling a macro looks a bit like a function call with an extra ! before its arguments. You've probably seen `vec![]`, which is indeed a macro. Other macros that are often used are `println!` and `dbg!`. When you're calling a macro, the ! at the end of the macro name is mandatory because it tells the compiler (and anyone reading the code) that you're trying to use a macro, not a regular function. A macro definition starts with `macro_rules!` followed by the name of the macro:

```
macro_rules! noop_macro {
    () => {};
}
```

The preceding macro does nothing. We can call it with `noop_macro!()`. You may notice that the body of the macro definition looks a bit like a `match` statement—because it *is* a `match` statement. You can match everything after the !, including different kinds of parentheses. You can use `()`, `{}`, or `[]`, but the parentheses are required.

Macros execute at compile time, so the code within them doesn't match the result of code execution, but rather the code itself. In other words, we can match the different kinds of code constructs in Rust, also known as *code fragments*. Here's another macro:

```
macro_rules! print_what_it_is {
    () => {                                    ← Matches no arguments,
        println!("A macro with no arguments")    such as print_what_it_is!()
    };
    ($e:expr) => {                             ← Matches one argument,
        println!("A macro with an expression")   provided that the argument
    };                                           is an expression such as
    ($s:stmt) => {                             ← print_what_it_is!({...})
        println!("A macro with a statement")
    };
}
        Matches one argument, provided
        that the argument is a statement,
        such as print_what_it_is!(...;)
```

The preceding macro has three matching rules: one that matches on no arguments, one that matches expressions, and one that matches statements. For the latter two rules, the arguments are available in the variables `$e` and `$s`, respectively. We can call our macro like so:

```
print_what_it_is!();
print_what_it_is!({});
print_what_it_is!(;);
```

Running this code produces the following output:

```
A macro with no arguments
A macro with an expression
A macro with a statement
```

Fragments can be any type of code construct in Rust so long as it's valid syntactically. You can also match multiple arguments:

```
macro_rules! print_what_it_is {
    // ... snip ...
    ($e:expr, $s:stmt) => {
        println!("An expression followed by a statement")
    };
}
```

If we call this macro with `print_what_it_is!({}, ;)`, it prints `"An expression followed by a statement"` when we run it. If we call the macro with an invalid argument (one that doesn't match any rules), we get a compiler error. Calling `print_what_it_is!` with two statements (`print_what_it_is!(;, ;)`) produces the following error:

```
error: no rules expected the token `,`
  --> src/main.rs:27:24
   |
5  | macro_rules! print_what_it_is {
   | -------------------------- when calling this macro
...
27 |     print_what_it_is!(;, ;); // error!
   |                        ^ no rules expected this token in macro call
```

We could write a match for this pattern, which would look like this:

```
macro_rules! print_what_it_is {
    // ... snip ...
    ($e:stmt, $s:stmt) => {
        println!("Two back-to-back statements")
    };
}
```

Although you could use any combination of statements and patterns as macro arguments, I implore you to think carefully about writing macros with too many argument combinations, as they can become quite confusing for the caller, particularly in complex cases. At the very least, it's important to understand the various scenarios so that you'll know what to do should you encounter them in the wild.

5.1.2   *When to use macros*

Macros look a lot like functions in Rust, so you might ask, "Why am I using a macro instead of a function?" Well, a couple of compelling use cases exist for using macros over functions. One reason why we might use macros is that macros allow us to overload arguments. Another reason is that macros support *variadic* arguments, which allow us to specify an arbitrary number of arguments with an optional separator. Other use cases for macros are custom logging (such as the `log` crate; https://crates.io/crates/log) and creating mini DSLs (as with the `lazy_static` crate; https://crates.io/crates/lazy_static).

Suppose that you want to write your own version of `println!`. The `println!` macro, as you may have noticed, takes N + 1 arguments (it is variadic). The first argument to `println!` is a string format specification, which may also contain interpolated variables, and the N arguments that follow are the values to be formatted. We can write our own macro that wraps `println!`:

```
macro_rules! special_println {              Matches any number
    ($($arg:tt)*) => {              ◄──     of token trees
        println!($($arg)*)          ◄───┐
    };                                  │   Passes the arguments
}                                       │   directly through to println!
```

We can call `special_println!` exactly as we would call `println!`. I simply copied the `println!` definition for the preceding example. Let's break down the argument specification, which looks like `$($arg:tt)*`:

- The identifier `$arg` is our named identifier for arguments that match this rule.
- We'll match on `tt`, which is short for *token trees*. A token tree can contain a single identifier, a sequence of identifiers, or a sequence of token trees (which in turn can contain identifiers because token trees are recursive). The recursive structure of token trees is why they're called *trees*.
- The match rule is within parentheses, as in `$(…)`, to denote the fact that the inner rule can be matched repeatedly. But we need to specify how many repetitions (as explained in the next item).
- The last character, an asterisk (`*`), tells the compiler that these arguments can repeat any number of times. Rust uses the same specs as regular expressions: `+` for one or more matches, `*` for any number of matches, and `?` for one or no matches.
- Because a token tree can be a sequence, we don't need to add punctuation because we're passing the entire token tree through.

The expansion or transcription of the arguments happens with `$($arg)*`, which we merely pass on to `println!`. Note that macros can call other macros, which also means that you can perform recursion within macros.

Let's make our macro slightly more useful. Suppose that we want to prefix all calls to our `special_println!` as we might do in a logging framework. Let's give that a shot:

```
macro_rules! special_println {                          We're passing all the
    ($($arg:tt)*) => {                                  arguments as the second
        println!("Printed specially: {}", $($arg)*)  ◄─ argument to println! In
    };                                                  the preceding example, we
}                                                       passed all the arguments
                                                        as the first argument.
```

Neat! Now if we call this code with `special_println!("hello world!")`, it prints
`"Printed specially: hello world!"`

Our macro has a problem, however: it accepts only one argument in its current
form, which isn't very useful. The reason is that we hardcoded the `{}` format specifier
as the first argument to our `println!` call, so `println!` expects and accepts only one
parameter when it's evaluated.

To make our macro accept any number of arguments (like `println!`), we can wrap
the arguments with `format!`, a special macro that correctly handles the string interpo-
lation and variadic arguments. The definition of the `format!` macro in the Rust standard library (https://doc.rust-lang.org/std/macro.format.html) looks a lot like our
code except that it calls the special `format_args!` macro and the `std::fmt::format()`
function from the Rust standard library.

> **Listing 5.1 `format!` macro definition from the Rust standard library**

```
macro_rules! format {
    ($($arg:tt)*) => {{
        let res = $crate::fmt::format(
          $crate::__export::format_args!($($arg)*));
        res
    }}
}
```

If we dig a little deeper, we see that `format_args!` is a special built-in macro (https://
mng.bz/ngaV) implemented by the compiler, so the buck stops there. (Rather, you'd
need to look into the Rust compiler source code to go deeper.) The following listing
shows the definition from the Rust standard library.

> **Listing 5.2 `format_args!` macro definition from the Rust standard library**

```
macro_rules! format_args {
    ($fmt:expr) => {{ /* compiler built-in */ }};
    ($fmt:expr, $($args:tt)*) => {{ /* compiler built-in */ }};
}
```

Moving on, let's update our `special_println!` to use `format!` to evaluate the arguments before the call to `println!`:

```
macro_rules! special_println {
    ($($arg:tt)*) => {
        println!("Printed specially: {}", format!($($arg)*))    ⟵
    };
}
```

**Our arguments are passed to format! before being
passed through to println! so that we also evaluate
the arguments as a formattable string.**

Now we can call the macro with `special_println!("with an argument of {}", 5)`,
and it will print `"Printed specially: with an argument of 5"`. To debug our macros,

we can enable the macro tracing feature (nightly only) by adding the following attribute:

```
#![feature(trace_macros)]
```

> **TIP** You can switch to nightly Rust by running `rustup default nightly` or overriding the tool chain for the current project with `rustup override set nightly`. You can also use the `+nightly` argument with `cargo` to run a specific crate with nightly Rust, like so: `cargo +nightly build`. Last, you can create a `rust-toolchain.toml` file in the root of your project with the following contents: `toolchain.channel = "nightly"`.

This code produces compiler messages that show the results of macro expansion. To use macro tracing, we need to enable it for specific invocations with `trace_macros!`:

```
trace_macros!(true);
special_println!("hello world!");
trace_macros!(false);
```

Now if we compile our code, it produces the following compiler output:

```
note: trace_macro
  --> src/main.rs:84:5
   |
84 |     special_println!("hello world!");
   |     ^^^^^^^^^^^^^^^^^^^^^^^^^^^^^^^^^
   |
   = note: expanding `special_println! { "hello world!" }`
   = note: to `println! ("Printed specially: {}", format! ("hello world!"))`
   = note: expanding `println! { "Printed specially: {}", format!
("hello world!") }`
   = note: to `{
             $crate :: io ::
             _print($crate :: format_args_nl!
             ("Printed specially: {}", format! ("hello world!"))) ;
         }`
   = note: expanding `format! { "hello world!" }`
   = note: to `{
           let res = $crate :: fmt ::
           format($crate :: __export :: format_args! ("hello world!")) ; res
       }`
```

Alternatively, we can use `cargo expand` to display the expanded macro, which is convenient when we want to stick with stable Rust. We can always run cargo with the `+nightly` argument to test our crate with nightly Rust and use nightly features.

> **TIP** You can install cargo-expand with `cargo install cargo-expand` if you've never used it before.

Let's write a new macro to demonstrate some more features. We'll write a macro that takes any number of identifiers and prints their values in the form `name=value`. This approach might be useful for debugging, for example. Keep in mind that `dbg!` already exists for this purpose, but we'll write our own macro as a learning exercise. The definition looks like this:

```
macro_rules! var_print {
    ($($v:ident),*) => {
        println!(
            concat!($(stringify!($v),"={:?} "),*), $($v),*
        )
    };
}
```

Matches a comma-separated list of identifiers

Stringifies and concatenates each argument as the first argument to println!, although including the full list of arguments as the remaining arguments to println!

This macro is more complicated, so let's break it down:

- Our macro matches on a comma-separated list of identifiers denoted by `$($v:ident),*`.
- The macro has two separate inner expansions of `$v`, one to produce the first argument for our call to `println!` and the other to pass the remaining arguments.
- The first argument for `println!` is a formatting string, which should contain each variable passed to the macro in the form `name=value`.
- The `stringify!` macro will convert the token to a string.
- The `concat!` macro will concatenate strings.
- The first expansion denoted by `$(stringify!($v),"={:?} "),*`: for each argument to the macro concatenates the stringified token with `"={:?}"`.
- The separate expansion is `$($v),*`, passed as the second argument to `println!`.
- Note that we're unwrapping the punctuation (`,`) in our match, so we have to add the punctuation back with `,*` in our expansion.

We can test our new macro as follows:

```
let counter = 7;
let gauge = core::f64::consts::PI;
let name = "Peter";
var_print!(counter, gauge, name);
```

Running this code produces the following output:

```
counter=7 gauge=3.141592653589793 name="Peter"
```

We can examine the expansion of this macro with `cargo expand`:

```
let counter = 7;
let gauge = 3.14;
let name = "Peter";
{
    ::std::io::_print(::core::fmt::Arguments::new_v1(
        &["counter=", " gauge=", " name=", " \n"],
        &[
            ::core::fmt::ArgumentV1::new_debug(&counter),
```

```
            ::core::fmt::ArgumentV1::new_debug(&gauge),
            ::core::fmt::ArgumentV1::new_debug(&name),
        ],
    ));
};
```

> **NOTE** You may notice that our `var_print!` macro is quite similar to the `dbg!` macro from the Rust standard library, although `dbg!` includes additional features. You may want to explore the `dbg!` macro definition to learn more.

Note that our code is further expanded by `println!`, which splits the format argument from one string into one for each argument. This process is handled internally by the compiler.

5.1.3 Using macros to write mini-DSLs

As I hinted in section 5.1.2, we can use macros to create miniature DSLs in Rust. The DSLs don't need to be miniature; they could be quite complex. But with macro-based DSLs, it's best to err on the side of simplicity.

The `lazy_static` crate (demonstrated in chapter 4) is a good example of using macros to make a DSL. Examine the macro definition in the following listing.

> **Listing 5.3 Macro definition from `lazy_static` crate**

```
macro_rules! lazy_static {
    ($(#[$attr:meta])* static ref $N:ident : $T:ty = $e:expr; $($t:tt)*) => {
        // use `()` to explicitly forward the information about private items
        __lazy_static_internal!($(#[$attr])* () static ref $N : $T =
$e; $($t)*);
    };
    ($(#[$attr:meta])* pub static ref $N:ident : $T:ty = $e:expr;
$($t:tt)*) => {
        __lazy_static_internal!($(#[$attr])* (pub) static ref $N : $T
= $e; $($t)*);
    };
    ($(#[$attr:meta])* pub ($($vis:tt)+) static ref $N:ident : $T:ty =
$e:expr; $($t:tt)*) => {
        __lazy_static_internal!($(#[$attr])* (pub ($($vis)+)) static ref
$N : $T = $e; $($t)*);
    };
    ()   ()
}
```

The macro looks complicated at first, but it's fairly simple. It matches only two possible patterns, which take the following forms:

- `static ref NAME: TYPE = EXPR;`
- `pub static ref NAME: TYPE = EXPR;`

The `$(#[$attr:meta])*` match at the start of each pattern allows you to include attributes (optional), and the `$($t:tt)*` at the end of each pattern makes the macro recursive by including everything after the first match in the `$t` variable.

The implementation details, such as code generation, are handled by the `__lazy_static_internal` macro, which uses recursion to expand the tail of the pattern, which is in the `$t` variable after the `;` that triggers the next recursion.

The last match, `() => ()`, provides a way to terminate the recursion when no more matches are found. Otherwise, an error would occur because on the next recursive match, the final expression would fail.

5.1.4   Using macros for DRY

Another common use case for declarative macros is for defining structures or code blocks that contain lots of repetition with minor variation. Sometimes, we want concrete implementations for many things that are identical except in terms of name or other properties. Typically, you'd use those macros privately; you likely wouldn't want to export them.

Suppose that we want to make a struct for each of the hundreds of dog breeds. Rather than define each struct separately, we can use a macro:

```
macro_rules! dog_struct {
    ($breed:ident) => {
        struct $breed {
            name: String,            We can store the
            age: i32,                name of the breed
            breed: String,      ◁─── within our structs.
        }
        impl $breed {
            fn new(name: &str, age: i32) -> Self {
                Self {
                    name: name.into(),
                    age,
                    breed: stringify!($breed).into(),   ◁─┐ We stringify the
                }                                          │ name and store
            }                                              │ it as a string.
        }
    };
}

dog_struct!(Labrador);
dog_struct!(Golden);
dog_struct!(Poodle);
```

Running `cargo expand`, we see the result of our `dog_struct!` macro:

```
struct Labrador {
    name: String,
    age: i32,
    breed: String,
}
impl Labrador {
    fn new(name: &str, age: i32) -> Self {
        Self {
            name: name.into(),
```

```
                age,
                breed: "Labrador".into(),
            }
        }
    }
    struct Golden {
        name: String,
        age: i32,
        breed: String,
    }
    impl Golden {
        fn new(name: &str, age: i32) -> Self {
            Self {
                name: name.into(),
                age,
                breed: "Golden".into(),
            }
        }
    }
    struct Poodle {
        name: String,
        age: i32,
        breed: String,
    }
    impl Poodle {
        fn new(name: &str, age: i32) -> Self {
            Self {
                name: name.into(),
                age,
                breed: "Poodle".into(),
            }
        }
    }
```

If we want to implement features of reflection in Rust, macros are one way to do so. We can't modify Rust code at run time, but we can emulate code creation at compile time with macros. We can add a trait to our dog structs to identify them:

```
trait Dog {
    fn name(&self) -> &String;
    fn age(&self) -> i32;
    fn breed(&self) -> &String;
}
```

Our Dog trait provides accessors for members of our breed structs and also allows us to identify dogs by using trait bounds as needed. Let's update our macro definition to use the trait:

```
macro_rules! dog_struct {
    ($breed:ident) => {
        struct $breed {
            name: String,
            age: i32,
```

```
            breed: String,
        }
        impl $breed {
            fn new(name: &str, age: i32) -> Self {
                Self {
                    name: name.into(),
                    age,
                    breed: stringify!($breed).into(),
                }
            }
        }
        impl Dog for $breed {
            fn name(&self) -> &String {
                &self.name
            }
            fn age(&self) -> i32 {
                self.age
            }
            fn breed(&self) -> &String {
                &self.breed
            }
        }
    };
}
```

We can test the reflection like this:

```
let peter = Poodle::new("Peter", 7);
println!(
    "{} is a {} of age {}",
    peter.name(),
    peter.breed(),
    peter.age()
);
```

The macro prints `"Peter is a Poodle of age 7"`.

Rust's declarative macros are quite powerful when used effectively. For problems of arbitrary complexity, you may need to use procedural macros (discussed in chapter 6). We can accomplish a lot with declarative macros, but they are somewhat limited in terms of what they can do.

Deciding whether to use macros is a matter of personal preference and coding style. In general, macros should be used sparingly and only when they offer clear advantages over alternative solutions. If you have code that is repetitive, verbose, or error-prone, it may be a good candidate for a macro. But macros should be used only when a significant portion of the code is repeated with only a few values, blocks, statements, or variables that need to be substituted. On the other hand, if your code is simple, clear, and easy to understand, it's probably better to leave it as is without using macros.

> **TIP** For a deeper discussion of Rust's macros and various ways to use them, see Sam Van Overmeire's book *Write Powerful Rust Macros* (Manning, 2024; https://www.manning.com/books/write-powerful-rust-macros). For reference documentation on Rust's declarative macro features, consult the Rust language reference on macros at https://mng.bz/v80m.

5.2 Optional function arguments

Many languages allow optional function arguments, but we can't use them in Rust. Optional function arguments allow you to specify default argument values in the function's definition or (in the case of languages such as C++ and Java) permit function overloading. Overloading is another way to express optional arguments, letting the compiler create distinct functions with the same name differentiated by the number or type of arguments. Both optional arguments and function overloading are forms of syntactic sugar.

Optional arguments are handy, allowing programmers to provide more flexibility to function callers. They're especially useful when you want to add new arguments to a function but retain backward compatibility.

Optional arguments aren't problem-free; they can lead to poor design when they're used excessively. Also, they encourage developers to reuse existing functions instead of creating new ones, which can make APIs more confusing. Finally, overuse of function overloading can make it hard to reason about what happens when you call a certain function, especially if the API changes over time.

5.2.1 Examining optional arguments in Python

So we can understand optional arguments better, let's see what they look like in another popular language: Python. An optional argument in Python would look something like this example—a function called `func` that accepts two arguments, each with a default value:

```
def func(optional_bool=True, optional_int=11):
    # ... function body goes here ...
```

Python's version of optional arguments is simple and succinct. We can specify default values right in the function definition for everyone to see, with little ambiguity. Python even allows us to specify each argument by name, not just position. We can call the function to specify only the second argument, as follows:

```
func(optional_int=1024)
```

Python's optional arguments are nice, but Rust takes a different approach, largely eschewing this style to preserve compatibility with C libraries.

5.2.2 *Examining optional arguments in C++*

C++ allows the use of optional function arguments through function overloading. That is, in C++ you can have multiple function definitions with different arguments, and the functions can supply a default for any missing arguments. This pattern in C++ with three overloaded functions might look something like this:

```
void func() {                          ←——|  Calls func() with
    func(true, 11);                        |  default values
}
void func(optional_bool: bool) {       ←——|  Calls func() with
    func(optional_bool, 11);               |  first default value
}
void func(optional_bool: bool, optional_int: int) {
    // ... function body goes here ...
}
```

C++ accomplishes this task by mangling function names, which makes C++ functions incompatible with C-based libraries. It's easy to call C code from C++, but calling C++ from C is best left unpursued.

5.2.3 *Optional arguments in Rust or the lack thereof*

Rust's explicit lack of optional arguments or overloading is a design choice, partly for C compatibility and partly to avert the criticisms mentioned in the preceding sections. We can emulate these features to varying degrees, however. We have three options:

- Extending with traits
- Using macros to match arguments at compile time
- Wrapping arguments with `Option`

We'll focus on the first pattern: extending with traits.

5.2.4 *Emulating optional arguments with traits*

First, we'll demonstrate that it's possible to have two traits with conflicted method names:

```
struct Container {
    name: String,
}
trait First {
    fn name(&self) {}
}
trait Second {
    fn name(&self) {}
}
impl First for Container {
    fn name(&self) {}
}
impl Second for Container {
    fn name(&self) {}
}
```

Here, we have two traits that differ only in name. Both traits are implemented for our `Container` struct. Everything looks good so far, but what would happen if we call `name()`? Let's try:

```
let container = Container {
    name: "Henry".into(),
};
container.name();
```

Compiling this code produces the following compiler error:

```
error[E0034]: multiple applicable items in scope
  --> src/main.rs:25:15
   |
25 |     container.name();
   |               ^^^^ multiple `name` found
   |
note: candidate #1 is defined in an impl of the trait `First` for the type
`Container`
  --> src/main.rs:14:5
   |
14 |     fn name(&self) {}
   |     ^^^^^^^^^^^^^^
note: candidate #2 is defined in an impl of the trait `Second` for the type
`Container`
  --> src/main.rs:18:5
   |
18 |     fn name(&self) {}
   |     ^^^^^^^^^^^^^^
help: disambiguate the associated function for candidate #1
   |
25 |     First::name(&container);
   |     ~~~~~~~~~~~~~~~~~~~~~~~
help: disambiguate the associated function for candidate #2
   |
25 |     Second::name(&container);
   |     ~~~~~~~~~~~~~~~~~~~~~~~~
```

This code makes complete sense. There's no way to disambiguate the function call. The compiler can provide helpful suggestions.

Next, what happens if the trait methods have different signatures? Let's add an argument to the `Second` trait (a `bool` parameter):

```
trait First {
    fn name(&self) {}
}
trait Second {
    fn name(&self, _: bool) {}
}
impl First for Container {
    fn name(&self) {}
}
```

```
impl Second for Container {
    fn name(&self, _: bool) {}
}
```

This code seems likely to work, but when you compile it, you get the same error. Let's try another way. We can use trait bounds by defining two functions like so:

```
fn get_name_from_first<T: First>(t: &T) {
    t.name()
}
fn get_name_from_second<T: Second>(t: &T) {
    t.name(true)
}
```

◀—| **Calls the name() from First, which takes only &self**

◀—| **Calls the name() from Second, which takes &self and a bool**

We can test it as follows:

```
let container = Container {
    name: "Henry".into(),
};
get_name_from_first(&container);
get_name_from_second(&container);
```

The compiler is happy. We've learned that we can use trait bounds to tell the compiler which method we want to use, depending on the context. Even when we have multiple conflicting traits, the compiler ignores traits that aren't specified in the trait bounds. If we have a generic function and try to call any method on a generic parameter, the compiler will complain:

```
fn get_name<T>(t: &T) {
    t.name()
}
```

This code errors out:

```
error[E0599]: no method named `name` found for reference `&T` in the
current scope
  --> src/main.rs:29:7
   |
29 |     t.name()
   |       ^^^^ method not found in `&T`
   |
   = help: items from traits can only be used if the type parameter is
   bounded by the trait
help: the following traits define an item `name`, perhaps you need to
restrict type parameter `T` with one of them:
   |
28 | fn get_name<T: First>(t: &T) {
   |              ~~~~~~~~~
```

```
28 | fn get_name<T: Second>(t: &T) {
   |               ~~~~~~~~~
```

What's neat about this example is that the compiler makes a good guess about what we're trying to do. With this knowledge, we can start thinking about optional arguments a bit differently. We know the following things:

- Function and method names cannot overlap even if their arguments are different.
- Traits may be implemented with conflicting methods for a type.
- If we use generics, we can specify trait bounds to disambiguate conflicting methods.

Thus, we can design our software to expect functionality delivered via traits. It's easy to use traits for this purpose in Rust because we can add trait bounds to any function. Except for base types such as `String` and numerics, it's often better to accept generic function parameters in Rust.

5.3   Builder pattern

The *builder pattern* is one of the original patterns described in the Gang of Four's *Design Patterns*. This pattern has become extremely popular in software design and (aside from iterators) is arguably one of the most enduring patterns from that book. The builder pattern can also be viewed as a form of *currying*, which is a way of converting a function that takes multiple arguments into a set of functions that take one argument each.

I'm a big fan of the builder pattern, and I consider it to be so useful that I included a whole section in this chapter for it. Implementing the builder pattern in Rust isn't particularly difficult, but we're going to work through an example in this chapter to tie together a lot of what we've learned in this book so far.

You might choose the builder pattern for multiple reasons, including encapsulation, convenience, separation of concerns, ergonomics, and safety. In Rust specifically, we normally don't want to expose structures directly, and as noted in section 5.2.4, Rust doesn't support optional arguments. So rather than rely on constructors with lots of arguments, we can use builders to handle more complex cases.

Builders aren't without problems; they add another layer of complexity. Knowing when to use them is more an art than a science.

5.3.1   Implementing the builder pattern

Let's write a basic builder for a bicycle we want to model. We're going to model the relationship shown in figure 5.1.

Now we'll implement the builder pattern.

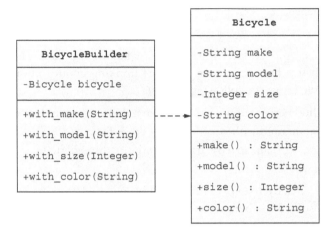

Figure 5.1   UML diagram for builder pattern

Listing 5.4   Code for builder pattern

```
#[derive(Debug)]
struct Bicycle {
    make: String,
    model: String,
    size: i32,
    color: String,
}

impl Bicycle {
    fn make(&self) -> &String {
        &self.make
    }
    fn model(&self) -> &String {
        &self.model
    }
    fn size(&self) -> i32 {
        self.size
    }
    fn color(&self) -> &String {
        &self.color
    }
}

struct BicycleBuilder {
    bicycle: Bicycle,
}

impl BicycleBuilder {
    fn new() -> Self {
        Self {
            bicycle: Bicycle {
                make: String::new(),
                model: String::new(),
                size: 0,
```

We're providing some accessors for the Bicycle struct.

Our BicycleBuilder struct holds a bicycle.

Constructing a BicycleBuilder will initialize our Bicycle with default values.

```
                color: String::new(),
            },
        }
    }
    fn with_make(&mut self, make: &str) {
        self.bicycle.make = make.into()
    }
    fn with_model(&mut self, model: &str) {
        self.bicycle.model = model.into()
    }
    fn with_size(&mut self, size: i32) {
        self.bicycle.size = size
    }
    fn with_color(&mut self, color: &str) {
        self.bicycle.color = color.into()
    }
    fn build(self) -> Bicycle {
        self.bicycle
    }
}
```

For each property of our Bicycle, we'll create a function to assign a value (such as a setter).

Calling build() will consume the builder and return the Bicycle by moving it out of the builder.

Our implementation satisfies the basic definition of a builder. Let's test it:

```
let mut bicycle_builder = BicycleBuilder::new();
bicycle_builder.with_make("Huffy");
bicycle_builder.with_model("Radio");
bicycle_builder.with_size(46);
bicycle_builder.with_color("red");
let bicycle = bicycle_builder.build();
println!("My new bike: {:#?}", bicycle);
```

Running the preceding code produces the following output:

```
My new bike: Bicycle {
    make: "Huffy",
    model: "Radio",
    size: 46,
    color: "red",
}
```

5.3.2 Enhancing our builder with traits

We can do some things to improve our implementation. We can start by creating a
Builder trait:

```
trait Builder<T> {
    fn new() -> Self;
    fn build(self) -> T;
}
```

We can rearrange our code for BicycleBuilder to implement the new trait:

```
impl Builder<Bicycle> for BicycleBuilder {
    fn new() -> Self {
```

```
        Self {
            bicycle: Bicycle {
                make: String::new(),
                model: String::new(),
                size: 0,
                color: String::new(),
            },
        }
    }
    fn build(self) -> Bicycle {
        self.bicycle
    }
}
```

While we're at it, we should add a trait to `Bicycle` that gives us an instance of the builder:

```
trait Buildable<Target, B: Builder<Target>> {
    fn builder() -> B;
}
```

Then we'll implement the `Buildable` trait for `Bicycle`:

```
impl Buildable<Bicycle, BicycleBuilder> for Bicycle {
    fn builder() -> BicycleBuilder {
        BicycleBuilder::new()
    }
}
```

Now we can get a new instance of our builder directly from a `Bicycle`:

```
let mut bicycle_builder = Bicycle::builder();
bicycle_builder.with_make("Huffy");
bicycle_builder.with_model("Radio");
bicycle_builder.with_size(46);
bicycle_builder.with_color("red");
let bicycle = bicycle_builder.build();
println!("My new bike: {:?}", bicycle);
```

Our code is starting to look more Rustaceous.

5.3.3 Enhancing our builder with macros

If we look at the `with_...()` methods in our builder, they look relatively redundant. Sometimes we want to specialize these functions, but generally, it's better to write a simple macro. Using a macro for lots of repetitive code is good because it helps us avoid typos. Let's give that approach a shot by replacing these methods with macros.

> **Listing 5.5 Adding `with_str!` and `with!` macros to `BicycleBuilder`**

```
macro_rules! with_str {
    ($name:ident, $func:ident) => {
```
with_str! accepts two idents: the member and function name.

```
        fn $func(&mut self, $name: &str) {
            self.bicycle.$name = $name.into()
        }
    };
}

macro_rules! with {
    ($name:ident, $func:ident, $type:ty) => {
        fn $func(&mut self, $name: $type) {
            self.bicycle.$name = $name
        }
    };
}

impl BicycleBuilder {
    with_str!(make, with_make);
    with_str!(model, with_model);
    with!(size, with_size, i32);
    with_str!(color, with_color);
}
```

The rendered function assigns the argument directly to the member, with a call to into() (from the Into trait).

The with! macro is nearly the same, except that it also accepts a type argument.

Listing 5.5 has two macros: `with_str!` and `with!`. The `with_str!` macro is for string fields, as we want to accept a `&str` for convenience, but we want to store the field as `String`. The `with!` macro accepts a type parameter, and we assume that the value is passed with a move. We could use a single macro to make the type optional, but the code is easier to understand this way.

> **TIP** Small one-off macros like those in this example are common. You can save yourself a lot of typing and errors by factoring out common parts into small reusable macros.

At this point, we can't do a lot more to improve our builder. We could make it a little more generic, but the returns are starting to diminish.

One thing we haven't discussed yet is visibility. We probably want to expose our types, traits, accessors, and builder methods, which we can do by adding the `pub` keyword as needed to `trait Buildable`, `Bicycle`, and `BicycleBuilder`. First, let's update the `Buildable` trait and `Bicycle` struct.

Listing 5.6   Public visibility for `Bicycle` and `Buildable`

```
pub trait Buildable<Target, B: Builder<Target>> {
    fn builder() -> B;
}

#[derive(Debug)]
pub struct Bicycle {
    make: String,
    model: String,
    size: i32,
    color: String,
}
```

Now the Buildable trait is public.

Now the Bicycle struct is public.

```
impl Buildable<Bicycle, BicycleBuilder> for Bicycle {
    fn builder() -> BicycleBuilder {
        BicycleBuilder::new()
    }
}
```

Next, let's add public visibility to the `Builder` trait and `BicycleBuilder`.

Listing 5.7 Public visibility for `Builder` and `BicycleBuilder`

```
pub trait Builder<T> {                    ◁────   Now the
    fn new() -> Self;                              Builder trait
    fn build(self) -> T;                           is public.
}

pub struct BicycleBuilder {               ◁────   Now the
    bicycle: Bicycle,                              BicycleBuilder
}                                                  struct is public.

impl Builder<Bicycle> for BicycleBuilder {
    fn new() -> Self {
        Self {
            bicycle: Bicycle {
                make: String::new(),
                model: String::new(),
                size: 0,
                color: String::new(),
            },
        }
    }
    fn build(self) -> Bicycle {
        self.bicycle
    }
}
```

We'll make one more tweak to our code by adding macros for the accessors. The final form of our builder macros looks like the following listing.

Listing 5.8 Final `Bicycle` and `BicycleBuilder` with macros

```
macro_rules! accessor {                       ◁────     We'll create one
    ($name:ident, &$ret:ty) => {         ◁────          accessor! macro with
        pub fn $name(&self) -> &$ret {                   two possible matches.
            &self.$name
        }                                            Matches on types where we
    };                                               want to return a reference
    ($name:ident, $ret:ty) => {              ◁────
        pub fn $name(&self) -> $ret {
            self.$name                           Matches on types where we
        }                                        want to return a copy (such
    };                                           as basic numeric types)
}
```

```
impl Bicycle {
    accessor!(make, &String);
    accessor!(model, &String);
    accessor!(size, i32);
    accessor!(color, &String);
}

macro_rules! with_str {
    ($name:ident, $func:ident) => {
        pub fn $func(&mut self, $name: &str) {
            self.bicycle.$name = $name.into()
        }
    };
}

macro_rules! with {
    ($name:ident, $func:ident, $type:ty) => {
        pub fn $func(&mut self, $name: $type) {
            self.bicycle.$name = $name
        }
    };
}

impl BicycleBuilder {
    with_str!(make, with_make);
    with_str!(model, with_model);
    with!(size, with_size, i32);
    with_str!(color, with_color);
}
```

NOTE Although creating these patterns can be a fun way to learn about the language and its features, much of this functionality is well covered by various crates. The `derive_builder` crate (https://crates.io/crates/derive_builder), for example, provides a way to create builders by using the `#[derive]` attribute. Although it's good to understand how to implement these patterns yourself, it's also good to know when to use existing solutions (such as `derive_builder`) to save time and benefit from the wisdom of crowds. The `derive_builder` crate in particular is full-featured, widely used, and battle-tested.

5.4 *Fluent interface pattern*

The *fluent interface* pattern builds on the builder pattern. The main characteristic that defines a fluent interface is the use of method chaining. *Method chaining* is the practice of chaining function calls together to perform an operation until the operation is terminated (usually by a method call that ends the operation).

We've already seen a good example of the fluent interface pattern in Rust: the `Iterator` trait. Method chaining can be accomplished by returning a type from each method call in the chain, which leads to the next step in the chain. The signature for the `map()` method on the `Iterator` trait looks like this:

```
fn map<B, F>(self, f: F) -> Map<Self, F> where
    F: FnMut(Self::Item) -> B { ... }
```

The return type here is Map, which is another iterator. We can call map() again, which will return another Map, and so on. Theoretically, we can chain functions infinitely this way.

5.4.1   A fluent builder

To demonstrate, let's revisit the builder example from the previous section. We'll update our assignment methods so that they return a builder. The updated Unified Modeling Language (UML) equivalent is shown in figure 5.2, in which each assignment method returns a new builder.

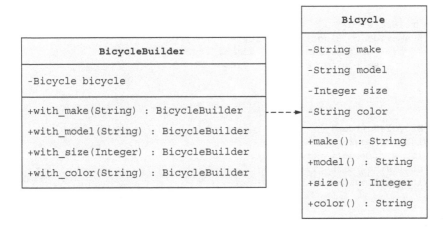

Figure 5.2   UML diagram for fluent builder pattern

Because we used macros, all we need to do is update the macros to implement this change:

```
macro_rules! with_str {
    ($name:ident, $func:ident) => {
        pub fn $func(self, $name: &str) -> Self {
            Self {
                bicycle: Bicycle {
                    $name: $name.into(),
                    ..self.bicycle
                },
            }
        }
    };
}

macro_rules! with {
    ($name:ident, $func:ident, $type:ty) => {
        pub fn $func(self, $name: $type) -> Self {
            Self {
```

```
                    bicycle: Bicycle {
                        $name,
                        ..self.bicycle
                    },
                }
            }
        };
}
```

Expanded, the code for our builder looks like this:

```rust
impl BicycleBuilder {
    pub fn with_make(self, make: &str) -> Self {
        Self {
            bicycle: Bicycle {
                make: make.into(),
                ..self.bicycle
            },
        }
    }
    pub fn with_model(self, model: &str) -> Self {
        Self {
            bicycle: Bicycle {
                model: model.into(),
                ..self.bicycle
            },
        }
    }
    pub fn with_size(self, size: i32) -> Self {
        Self {
            bicycle: Bicycle {
                size,
                ..self.bicycle
            },
        }
    }
    pub fn with_color(self, color: &str) -> Self {
        Self {
            bicycle: Bicycle {
                color: color.into(),
                ..self.bicycle
            },
        }
    }
}
```

Cool! Notice a couple of things in the preceding code:

- Our assignment methods take `self`, not `&mut self`. In other words, each call to an assignment method consumes the previous builder.
- Rather than copy or return the old builder and inner struct, we'll create a new builder with a new inner struct.

- We're using the spread syntax (..) to initialize the `Bicycle` struct with our updated field.

Initializing structs with the spread syntax

If you've never seen the spread syntax on struct initialization, don't be alarmed. This notation is handy for initializing a struct with the values of an existing struct while updating specific fields. The operation uses a move, so it consumes the existing struct upon assignment. The spread syntax is syntactic sugar to make handling structs with many fields easier.

One handy side effect of the spread syntax is that it allows us to change fields in a struct even when it's not mutable, but only if it's owned. Our `Bicycle` struct demonstrates this concept:

```
let bicycle1 = Bicycle {                    ◁──┐  We create a new instance of
    make: "Rivendell".into(),                   │  our Bicycle struct with all
    model: "A. Homer Hilsen".into(),            │  fields specified.
    size: 51,
    color: "red".into(),
};
println!("{:?}", bicycle1);
let bicycle2 = Bicycle {        ◁──┐  We create a new instance of the
    size: 58,                       │  same struct, but we've changed the
    ..bicycle1                      │  size field to a different value.
};
println!("{:?}", bicycle2);
// println!("{:?}", bicycle1);  ◁──┐  We can't use bicycle1 after using
                                     the spread syntax because it gets
                                     moved into the new struct.
```

Running the preceding code produces the following output:

```
Bicycle { make: "Rivendell", model: "A. Homer Hilsen", size: 51,
color: "red" }
Bicycle { make: "Rivendell", model: "A. Homer Hilsen", size: 58,
color: "red" }
```

Because the assignment uses a move, you can't use the spread syntax with references. Trying to compile the following code will produce an error:

```
let bicycle = Bicycle {
    make: "Rivendell".into(),
    model: "A. Homer Hilsen".into(),
    size: 51,
    color: "red".into(),
};
let bicycle = Bicycle {            │  Compiler produces an error with
    size: 58,                      │  "mismatched types expected struct
    ..&bicycle        ◁────────────┤  Bicycle, found &Bicycle ".
};
```

5.4.2   Test-driving our fluent builder

Let's update our test code to use the new fluent interface. Our updated code looks like this:

```
let bicycle = Bicycle::builder()
    .with_make("Trek")
    .with_model("Madone")
    .with_size(52)
    .with_color("purple")
    .build();
println!("{:?}", bicycle);
```

Neat! That looks much better than the old form.

5.5   Observer pattern

The *observer pattern* (along with its variations) is widely used to enable objects to observe changes in other objects. Observer is one of the patterns from *Design Patterns* and is often necessary in systems that perform any kind of event processing or event handling, such as networked services.

5.5.1   Why not callbacks?

Before we dive deeper into the observer pattern, let's discuss callbacks. Some languages (notably JavaScript) make heavy use of callbacks, which can lead to a situation known as *callback hell*, with deeply nested callbacks within callbacks creating difficult-to-understand code. Someone went so far as to create the website http://callbackhell.com to describe this problem and propose some solutions.

Callbacks are often used in functional languages within higher-order functions. A *higher-order function* is a function that takes another function as a parameter or returns another function. Iterators use callbacks with functions like map(), for example. The basic form of a callback in Rust looks something like this:

```
fn callback_fn<F>(f: F)
where
    F: Fn() -> (),
{
    f();
}

fn main() {
    let my_callback = || println!("I have been called back");
    callback_fn(my_callback);
}
```

Nothing happens here. Our function hasn't been called—only declared.

Our callback is called within callback_fn().

In the preceding example, I'm using a closure for the callback (which is what you typically see in JavaScript), but I could just as easily pass an ordinary function. Simple cases like this example are fine, but following the logical flow can get messy and confusing when you have callbacks within callbacks within callbacks.

Although callbacks are not necessarily bad on their own, the observer pattern provides looser coupling, makes it easier to attach and detach observers (equivalent callbacks), and enables us to have a many-to-one relationship instead of one-to-one. More generally, we can use the observer pattern when we have code that needs to notify other code about events (the subject) without needing a dependency on the observers. The subject can notify the observers without needing to know the observers.

One more problem with callbacks is that they don't allow us to decouple state from the function we pass to the callback. We have to bind the state to a callback by using a closure or global variables.

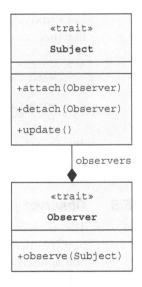

Figure 5.3   UML diagram of the observer pattern

5.5.2   *Implementing an observer*

There are multiple ways to implement the observer pattern, and each way has tradeoffs. The example in this section is flexible enough that we can change the implementation details as needed to suit a variety of cases. We're going to implement the observer pattern in Rust as shown in figure 5.3.

We'll start by implementing two traits: `Observer` and `Observable`. We'll use `Observer` for objects that want to observe others. `Observable` will be implemented by objects that want to allow other objects to observe them. The following listing shows the `Observer` trait.

Listing 5.9   `Observer` trait

```
pub trait Observer {
    type Subject;                                      ⟵ We use an associated
                                                            type for the subject.
    fn observe(&self, subject: &Self::Subject);   ⟵ The observe() method is
}                                                        called by the subject when
                                                          an update occurs.
```

For the observer, I'm using the term *observe* instead of *notify* (per the original design pattern). Next, consider the following listing, which shows the `Observable` trait.

Listing 5.10   `Observable` trait

```
pub trait Observable {
    type Observer;                                     ⟵ An associated type is
    fn update(&self);                                       used for the observer.
    fn attach(&mut self, observer: Self::Observer);
    fn detach(&mut self, observer: Self::Observer);
}
```

The `Observable` trait provides the methods for our subject and matches the original design pattern. We don't make any assumptions about the type of the observer or the subject, which gives us a bit more flexibility with this pattern. Next, we need to create a subject and implement `Observable` on it.

Listing 5.11   Implementing `Observable` for `Subject`

The observer should be provided as an Arc, which provides additional flexibility and shared ownership.

We're storing weak pointers to our Observer objects, where the subject is Self.

```
pub struct Subject {
    observers: Vec<Weak<dyn Observer<Subject = Self>>>,
}

impl Observable for Subject {
    type Observer = Arc<dyn Observer<Subject = Self>>;
    fn update(&self) {
        self.observers
            .iter()
            .flat_map(|o| o.upgrade())
            .for_each(|o| o.observe(self));
    }
    fn attach(&mut self, observer: Self::Observer) {
        self.observers.push(Arc::downgrade(&observer));
    }
    fn detach(&mut self, observer: Self::Observer) {
        self.observers
            .retain(|f| {
                !f.ptr_eq(&Arc::downgrade(&observer))
            });
    }
}
```

self.observers holds weak references, which need to be upgraded here. Because upgrade() on Weak returns an option, flat_map() unwraps and removes the None cases.

Last, we call observe() on each observer that's still valid.

When a new observer is added, we downgrade it from an Arc to a Weak pointer.

We have to use ptr_eq() to find the matching object. Vec::retain() filters out all the objects that match the pointer passed to this method.

I chose to require that observers be passed as `Arc<dyn Observer>`, which provides a bit of additional flexibility. For one thing, we can store the pointers as weak pointers, which means that when they go out of scope, we can ignore them instead of keeping the object alive. Using `Arc` also allows shared ownership (that is, we don't want our subject to take ownership of the observers). Because the observer is defined as an associated type in the trait, we could easily change the type from `Arc` to something else while reusing the same traits.

Next, let's add some state to our subject so that we can test it, provide an accessor, and add a `new()` method. We'll update the code so that it looks like the following listing.

Listing 5.12   Adding state and `new()` to `Subject`

```
pub struct Subject {
    observers: Vec<Weak<dyn Observer<Subject = Self>>>,
    state: String,
}

impl Subject {
    pub fn new(state: &str) -> Self {
```

```
        Self {
            observers: vec![],
            state: state.into(),
        }
    }

    pub fn state(&self) -> &str {
        self.state.as_ref()
    }
}
```

Next, let's create an observer and implement the `Observer` trait for it.

Listing 5.13   Creating an observer

```
struct MyObserver {
    name: String,                    ◁───┤  We add a name to our observer
}                                          so we can identify it.

impl MyObserver {
    fn new(name: &str) -> Arc<Self> {      ◁───┤  Our new() method will return
        Arc::new(Self { name: name.into() })      an Arc<Self> instead of Self.
    }
}
                                                    Our subject type is
                                                    Subject, which we
impl Observer for MyObserver {                      defined in listing 5.12.
    type Subject = Subject;                ◁───┘
    fn observe(&self, subject: &Self::Subject) {    ◁───┤  Our observe()
        println!(                                           implementation prints
            "observed subject with state={:?} in {}",       the state from our
            subject.state(),                                subject and the name of
            self.name                                       this observer instance
        );                                                  when called.
    }
}
```

Finally, we can test our observer.

Listing 5.14   Testing the observer pattern

```
let mut subject = Subject::new("some subject state");

let observer1 = MyObserver::new("observer1");
let observer2 = MyObserver::new("observer2");

subject.attach(observer1.clone());        │  We have to clone our pointer; otherwise, it
subject.attach(observer2.clone());        │  will go out of scope when passed by value.

// ... do something here ...
                                          │  Normally, we'd call update() from within the subject
subject.update();              ◁───┤         whenever its state changes to trigger our observers,
                                             but for example purposes, we call it here.
```

Running this code produces the following output:

```
observed subject with state="some subject state" in observer1
observed subject with state="some subject state" in observer2
```

5.6   Command pattern

The *command pattern* stores state or instructions in one structure and
then applies changes later. This pattern is widely used but not well
specified and arguably a bit dated. For the sake of completeness,
however, I'll document a simple example of implementing the command pattern in Rust.

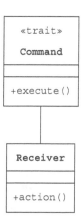

5.6.1   Defining the command pattern

Before we implement this pattern in Rust, let's define the essence of
the command pattern. We need to concern ourselves mainly with a
single trait called `Command` that executes against a `Receiver`. The
`Receiver` should be a concrete object of some kind, though it doesn't
necessarily need to be called `Receiver` or have a method called
`action()`. This pattern resembles the object-oriented version
described in the Gang of Four's book *Design Patterns*. Figure 5.4 illustrates the relationship between these traits.

**Figure 5.4
UML diagram for
the command
pattern**

We'll define our `Command` trait as shown in the following listing.

Listing 5.15   Command trait definition

```
trait Command {
    fn execute(&self) -> Result<(), Error>;
}
```

Note that I've made the `Command` trait return a result, which provides some basic error
handling (as we'll see in section 5.6.2). The `Command` trait is the essence of the command pattern, but we need to put all the pieces together for this pattern to make
sense. We have to supply a receiver, which is any object on which the command can
execute, as defined by the concrete implementation of the `Command` trait.

5.6.2   Implementing the command pattern

For this example, we're going to create two command objects that operate on file handles: a command to read a file and a command to write a file. The receiver will be a
file handle. First, let's define our `ReadFile` command.

Listing 5.16   `ReadFile` command implementation

```
struct ReadFile {
    receiver: File,          ◁──┤  receiver is the receiver (or
}                                   target) of our command.
```

```
impl ReadFile {
    fn new(receiver: File) -> Box<Self> {
        Box::new(Self { receiver })
    }
}

impl Command for ReadFile {
    fn execute(&self) -> Result<(), Error> {
        println!("Reading from start of file");
        let mut reader = BufReader::new(&self.receiver);
        reader.seek(std::io::SeekFrom::Start(0))?;

        for (count, line) in reader.lines().enumerate() {
            println!("{:2}: {}", count + 1, line?);
        }

        Ok(())
    }
}
```

We're returning a boxed object so we can use trait objects later . We don't have to return a Box at this point, but it makes the code a bit cleaner and tells the caller how we expect the code to be used.

Each line is printed, and if an error occurs while reading the file, we'll return it via the ? operator.

We'll use a buffered reader, which gives us a few nice features such as an iterator over lines in file.

We enumerate the lines so that we can print the line number along with its content.

We always seek to the start of the file before reading it back. We use the ? operator to handle I/O errors gracefully.

Note that we implement a new() method for ReadFile, which takes in a file handle and returns a boxed ReadFile object. This process is important, as we'll see in listing 5.18. Next, let's define the WriteFile command.

Listing 5.17 WriteFile command implementation

```
struct WriteFile {
    content: String,
    receiver: File,
}

impl WriteFile {
    fn new(content: String, receiver: File) -> Box<Self> {
        Box::new(Self { content, receiver })
    }
}

impl Command for WriteFile {
    fn execute(&self) -> Result<(), Error> {
        println!("Writing new content to file");
        let mut writer = self.receiver.try_clone()?;
```

The command includes a content field, which is the content we want to write to the file.

receiver is the file handle to which the command applies, as with ReadFile.

As with ReadFile, we return a boxed object.

We need a mutable object to write to the file, and the easiest way to obtain one is to clone our file handle.

```
writer.write_all(self.content.as_bytes())?;
writer.flush()?;

        Ok(())
    }
}
```

We convert the UTF-8 string to raw bytes and write all the bytes to the current location of the file cursor.

We flush the file handle to make sure that the bytes are written out, handling errors via the ? operator.

The `WriteFile` command looks quite similar to `ReadFile` except that we also include the contents we want to append to the file as an argument. Note that in our implementation, we assume that `WriteFile` can happily write to the current file-handle position, but a slightly more robust implementation might always seek to the end of the file before writing (and we might call it `AppendFile` instead). I'll leave this change as an exercise for you.

Next, we need to implement the client of the pattern. We'll stick the client in our `main()` method.

Listing 5.18   Client implementation for command pattern

```
use std::fs::File;
use std::io::{BufRead, BufReader, Error, Seek, Write};

fn main() -> Result<(), Error> {
    let file = File::options()
        .read(true)
        .write(true)
        .create(true)
        .append(true)
        .open("file.txt")?;

    let commands: Vec<Box<dyn Command>> = vec![
        ReadFile::new(file.try_clone()?),
        WriteFile::new(
          "file content\n".into(), file.try_clone()?
        ),
        ReadFile::new(file.try_clone()?),
    ];

    for command in commands {
        command.execute()?;
    }

    Ok(())
}
```

We use std::fs::File from the standard library to open a file in read/write mode and create the file if it doesn't exist or open it in append mode if it does.

We use trait objects with Box<dyn Command>, which allows us to put any commands that implement the Command trait in our list of commands.

When we create each command, we clone the file handle and pass it along to the command.

Note that we need to include the newline \n character at the end of the file contents to give us separate lines in the text file.

We loop over each command object and call its execute() method, using the ? operator to handle errors.

NOTE   We introduced trait objects (such as the use of `dyn Trait`) in chapter 2 when discussing traits.

Now we can test our code with `cargo run`, which (provided that `file.txt` doesn't exist) will produce the following output:

```
Reading from start of file
Writing new content to file
Reading from start of file
 1: file content
```

Because this example is stateful, we'll see a different result when we run the code a second time because the file was modified in a previous run. Running the code a second time produces the following output:

```
Reading from start of file
 1: file content
Writing new content to file
Reading from start of file
 1: file content
 2: file content
```

The command pattern can be much more complicated because it is sometimes used for stateful operations, such as those that can be applied in forward and reverse. An example might be an undo/redo framework, which permits us to apply and reverse changes. For this approach to work, our commands need to be idempotent and track the necessary state for both forward and reverse execution. In the preceding example, the `WriteFile` command is not idempotent because it performs an append operation and doesn't seek to the end of the file each time. One way to make the command idempotent would be to seek to the beginning of the file each time and overwrite the entire contents of the file.

5.7   *Newtype pattern*

The *newtype pattern* is an extension of *tuple structs* (special structs in Rust that behave like tuples) that uses Rust's type system to provide additional type information or handling of data. Newtype is useful when the data itself is sufficiently contained by a core or primitive type, such as a `String` or `i32`. But you want to avoid adding too much encapsulation or indirection on top of the base type by allowing direct access with a tuple.

We can think of newtype as being a lightweight pattern for providing additional context or information atop tuples while keeping the convenience and simplicity of tuples. Another common use of newtype is enabling type-safe conversion between data types. Newtype can be deceptively simple, but it's also deceptively handy for using Rust's type system.

> **NOTE** In the introduction to this chapter, I noted that newtype is a *Rust-specific* pattern, though strictly speaking, nothing stops you from applying the same concept in another programming language. By *Rust-specific*, I mean merely that this pattern (to the best of my knowledge) originated within the Rust community.

To demonstrate the use of newtype, let's create `BitCount` and `ByteCount` types, which we'll use to hold counts of bits and bytes. We know that 1 byte contains 8 bits, so we can define methods to trivially (but safely) convert between these types. First, we'll define our tuple structs, each with a `u32`:

```
#[derive(Debug)]
struct BitCount(u32);
#[derive(Debug)]
struct ByteCount(u32);
```

This code represents the most basic example of the newtype pattern. We can test it like so:

```
let bits = BitCount(8);
let bytes = ByteCount(12);
dbg!(&bits);
dbg!(&bytes);
```

Running this code produces output like this:

```
[src/main.rs:9] &bits = BitCount(
    8,
)
[src/main.rs:10] &bytes = ByteCount(
    12,
)
```

Next, we want to convert between counts of bits and bytes. Let's define two methods to perform this conversion:

```
impl BitCount {
    fn to_bytes(&self) -> ByteCount {
        ByteCount(self.0 / 8)          ◁──┐  May return an
    }                                      │  unexpected result if the
}                                          │  number of bits is not
                                           │  evenly divisible by 8
impl ByteCount {
    fn to_bits(&self) -> BitCount {
        BitCount(self.0 * 8)
    }
}
```

Conversion method naming idioms: as_...(), to_...(), and into()

You may have noticed that three common idioms are used for method names when converting between types: prefixing methods with `as_` or `to_` and `into()`. Although developers don't strictly follow these conventions, you'll find that most libraries (especially the Rust standard library) adhere to the following conventions:

- `as_...()`—For lower-cost conversions such as `as_ref()` from the `AsRef` trait. Obtaining a reference is a relatively cheap operation—one that in some cases can be optimized out by the compiler.

(continued)

- `to_...()`—For higher-cost conversions such as `to_string()` from `ToString`. The imperative `to` implies that work needs to be done, such as allocating, creating new objects, performing conversions, or copying data.
- `into()`—Conversions using `into()` (via the `From` trait). These conversions are generally higher cost and often include allocations, copying, or cloning.

One notable exception is the use of `borrow()` from the `Borrow` trait, which behaves similarly to `as_ref()` from `AsRef` except that it returns a reference object (a pattern we'll discuss in chapter 7) rather than a plain reference (`Ref<', T>` versus `&T`). `std::cell::RefCell`, for example, provides `borrow()` but not `as_ref()` because of the additional overhead introduced by run-time borrow checking.

We can check whether our conversions behave as expected with the following code:

```
dbg!(bits.to_bytes());
dbg!(bytes.to_bits());
```

When we execute this code, we get the following output, which shows the new object produced:

```
[src/main.rs:24] bits.to_bytes() = ByteCount(
    1,
)
[src/main.rs:25] bytes.to_bits() = BitCount(
    96,
)
```

We can convert from bits to bytes and back again, and vice versa, if we're so inclined:

```
dbg!(bits.to_bytes().to_bits());
dbg!(bytes.to_bits().to_bytes());
```

Running this code produces the following output:

```
[src/main.rs:27:5] bits.to_bytes().to_bits() = BitCount(
    8,
)
[src/main.rs:28:5] bytes.to_bits().to_bytes() = ByteCount(
    12,
)
```

Accessing the inner value of a newtype is as simple as using the tuple syntax, as newtypes are in effect tuples:

```
dbg!(bits.0);
dbg!(bytes.0);
```

The preceding code produces the following output:

```
src/main.rs:30:5] bits.0 = 8
[src/main.rs:31:5] bytes.0 = 12
```

Converting between units—such as bits and bytes, Celsius and Fahrenheit, and meters and feet—is a common use case for newtype, as this pattern allows you to encode the conversion logic in a single place and ensures that the conversion is always correct. Note that if your conversion involves a lossy operation such as floating-point math, you could lose precision with each conversion, so you may want to consider keeping the source value around for future conversions.

Newtype is convenient, doesn't require much boilerplate, and is fairly easy for other people to grok. The pattern is essentially named tuples with one or more methods defined, such as for converting between related types.

Summary

- Macros provide one method of metaprogramming in Rust. We can generate code with macros, saving ourselves a lot of typing and reducing the number of errors that can appear when we need to generate or create repetitive code.
- The core language patterns of Rust (generics and traits) can be used to create advanced patterns such as builder and fluent interface.
- The builder pattern demonstrates how to use encapsulated data effectively and separate concerns.
- The fluent interface pattern is a pleasant way to deal with chaining operations and converting between types.
- The observer pattern is an alternative to callbacks, providing a cleaner abstraction at the expense of some boilerplate. For simple cases, callbacks may be sufficient.
- The command pattern gives us a method to abstract the execution of a command from the target (or receiver) of the action, as well as the order and timing of execution.
- The newtype pattern wraps other types within a tuple struct to encode additional information about a type or enable safe data conversions. Core types such as `String` or primitives such as `i32` are candidates for newtype. Newtype allows us to convert easily between similar but distinct types.

Designing a library 6

This chapter covers
- Thinking about how to design a great library
- Making beautiful interfaces
- Being correct and avoiding unexpected behavior
- Exploring Rust library ergonomics and patterns

This chapter marks the approximate halfway point in this book, so we'll take a slight departure from the other content to discuss a subjective and somewhat controversial subject: what constitutes good library design. There's no controversy about good design being better than bad design, but few people agree on what constitutes *good*. The zeitgeist of opinion surrounding good versus bad tends to shift and swing over time, which is important to consider for your designs.

The truth about good software design is that few universal rules exist. Much of what constitutes good is a matter of fashion, context, availability, and quality of tooling, as well as how the human–computer interface functions across these dimensions. That interface is the API of your library, and it's the most important part of your library design.

In this chapter, we'll explore some of the ideas, processes, and methods to consider when designing a library, with the goal of producing a library that's easy to use, delightful to work with, difficult to use incorrectly, and flexible enough to solve a wide variety of problems. We'll use an example from earlier in the book to build our library.

This chapter is written for people who are interested in publishing their own libraries as open source projects, SDKs, or APIs for internal use. It's also for those who want to learn more about the process of designing libraries and the considerations that go into making a library that's easy to use, maintain, and extend.

Before diving into a specific example, we'll take a moment to contemplate some of the problems we face as custodians of software libraries. This meditation will set the stage for a more practical example.

6.1 Meditate on good library design

Designing a library—or anything, for that matter—always involves tradeoffs. We can think of tradeoffs as being a sliding scale; every choice we make as software developers is about striking the right balance among tradeoffs, which can be binary, scalar, 3D, 4D, or N-dimensional. Somewhere along the continuum of these tradeoffs is a point that represents a good balance.

An example binary choice might be whether to add a dependency to implement a feature or write a solution yourself. Scalar decisions involve striking a balance between at least two options, such as configuration versus convention (making everything configurable, some things configurable, or nothing configurable).

For most practical programming tasks, the main constraint is delivering the necessary features in minimum time without sacrificing quality. Across the three dimensions of speed, completeness, and quality, you'll likely need to compromise on one or more dimensions to optimize for the others (sacrificing speed for quality, for example, or dropping some features to allow shipping sooner).

When it comes to designing library APIs, we can look to Marie Kondo for inspiration. We want our library to spark joy in those who use it, and we need to get inside the heads of our library's audience to understand what's joyful and what is not. In many cases, this process is as easy as using your library, comparing its interface with that of similar or related libraries, and ensuring that the interface and patterns your library exposes are congruent with what people expect to find in a library. We should trim the interfaces exposed by our library that don't spark joy.

6.2 Do one thing, do it well, and do it correctly

As good stewards of the Rust ecosystem, we want to produce libraries that are Rustaceous and that follow the Rust ethos. Many crates focus on doing a small set of things and doing those things well. We want our crates to be interoperable with other crates. We don't want to pull in too many dependencies, and when we do impose dependencies on downstream consumers, we want to make sure that we don't break things. Sometimes, we make dependencies or features optional by using feature flags, but too many feature flags can be confusing and make a library harder to use. Trying to achieve all these aims simultaneously is a tricky balancing act that becomes much harder as the complexity of the library increases.

Being good at one thing is a good way to ensure that your library is easy to test, easy to maintain, and easy to use. It's also a good way to ensure that your library is correct.

Correctness is more important than performance or completeness. Achieving correctness is a matter of ensuring that your library does what it says (that is, matches the specifications or documentation) and does it predictably and reliably. It's harder to be correct when you're trying to do too many things or when you're doing things in a way that is not idiomatic to the language you're using and the context in which it's used.

Proving correctness is a complex topic that can't be summarized in a single chapter, but we can use tools such as property-based testing, fuzzing, and formal verification to ensure that our libraries are correct. Formal verification, in particular, is the hardcore end of the spectrum; it's not something that most of us will ever need to do, but it's good to know that it's possible. Property-based testing and fuzzing are more accessible and can be used to great effect to ensure that our libraries are correct.

6.3   *Avoid excessive abstraction*

As library designers, we need to decide what to expose on public interfaces. In most cases, we start by exposing the minimum set of types, methods, traits, and functions that provide minimally necessary feature completeness. We don't want to use excessive abstractions or encapsulation, particularly for raw data; instead, we want to empower the downstream consumers of our library to handle data as they see fit. We'll implement common traits (`Debug`, `Clone`, and so on) to make life easy, but we don't need to follow the kitchen-sink approach and implement every trait simply because we can.

The downside of too much abstraction is that it can make your library harder to use, raising the barrier to entry and discouraging people from using it, especially when the abstractions your library introduces are not idiomatic to the language or the problem domain and differ from what people expect to find in a library. If the abstractions are too complex, they can make your library incompatible with other libraries, which is a problem if you want the library to be used in various contexts.

As the old joke says, when Michelangelo was asked how he created the statue of David, he replied, "All I did was chip away everything that didn't look like David." The same is true of library design. We should chip away at the abstractions that aren't necessary until we're left with the simplest, most elegant solution to the problem we need to solve.

6.4   *Stick to basic types*

One way to ensure that your library is accessible to a wide range of applications is to stick to basic types when possible. Introducing new types and custom data structures means that anyone else who uses your library needs to take an extra step to convert between their data structures and yours.

Ideally, you can rely entirely on the standard library types, and if you need to introduce new types, you should make sure that they are easy to convert to and from standard library types, providing the necessary conversions (such as implementing `From`). Also, requiring users to convert between types introduces some performance overhead, which may be undesirable.

Rust's standard library and collections (including `Vec`, `HashMap`, and `HashSet`) are quite suitable for most tasks, and you should consider using them whenever possible. But you can go further by accepting slices or iterators as input to your functions, making your library even more flexible.

Consider a library that accepts a `Vec` as input. This interface is less flexible because we can pass only a `Vec`:

```
fn do_something_with_vec<T>(v: &Vec<T>) {
    // ...
}
```

This interface is more flexible because we can pass a `Vec`, an array, or any other type to a slice:

```
fn do_something_with_slice<T>(v: &[T]) {
    // ...
}
```

A slice may be slightly less flexible than an iterator, but it's quite a bit more flexible than a `Vec`.

6.5 Use the tools

Tools such as Clippy and rustfmt can enforce compliance with Rust's idioms and conventions. It's Rustaceous, for example, to use camel case for types, snake case for variables or member functions, upper case for constants, and so on; Clippy provides lints for all these conventions. Clippy is one of the most helpful tools for ensuring that your code is idiomatic Rust.

Clippy and rustfmt relate mainly to idioms, so they can't do much to help with design, architecture, or correctness. But they can help you avoid common pitfalls and ensure that your code is relatively easy to read and understand.

Integrating Clippy and rustfmt into your editor and continuous integration/continuous delivery (CI/CD) pipeline is an easy way to ensure that code stays compliant over time. The cost of changing code after it's written is much higher than the cost of writing code correctly in the first place, so these tools are worth using, especially because they're free, easy to use, and trivial to integrate.

6.6   *Good artists copy; great artists steal (from the standard library)*

When you are unsure which conventions to follow, popular Rust crates can serve as data points you can analyze to understand what works and what doesn't. Following the lead of popular crates usually helps you avoid bad designs. This statement isn't necessarily an endorsement of any crate that's managed to achieve popularity, however.

When you're seeking inspiration, the Rust standard library is the gold standard for idiomatic Rust. The standard library is well documented, well tested, and well designed. You can examine the source code and historical discussions in the Rust repository to understand why the language's developers made specific decisions.

The official documentation for the standard library links directly to the source code, which is a good resource for understanding how things work. The Rust language and its standard library are dual-licensed under Apache 2.0 and Massachusetts Institute of Technology (MIT), so in most cases, you can use examples from Rust's source code in your projects as a starting point.

6.7   *Document everything, and provide examples*

Documenting your library is a critical step in the process; you shouldn't think of it as being a stage at the end of writing your code. Instead, you should create the documentation, including example code, throughout the process of writing your library.

Examples are sometimes overlooked but are some of the most important parts of documentation. Typically, someone who uses your library begins by copying and pasting an example from the documentation and modifying it to suit their needs. I imagine that anyone who has used a library has done the same thing at least once and is nodding in agreement as they read.

6.8   *Don't break the user's code*

We should make an effort to maintain backward compatibility whenever possible. For crates that we publish, we should use semantic versioning to signal compatibility between versions to our downstream consumers. If we want to publish our crate, maintaining our library is an ongoing process that requires fluidity in terms of adopting new features and patterns and eschewing those that have gone out of style.

Backward compatibility is such a precious trait in a library that you should go out of your way to maintain it. It's better to have a slightly less optimal API than to break your users' code. If you must break compatibility, you should provide a migration path for your users and communicate the changes clearly in your documentation and release notes.

Bear in mind that when developers make backward-incompatible changes, many folks don't bother reading the release notes or the change logs or checking documentation. They'll simply update their dependencies and expect everything to work. As library maintainers, maintaining backward compatibility will spare us and our library's users a lot of headaches.

6.9 *Think of the state*

One of the most critical aspects of designing a library is thinking about the way we want users of our libraries to handle state. A few things we probably shouldn't do are create global variables and use mutable statics and singletons.

Most good library designs provide a way for users to create instances of the context in which their library operates. That context in turn is passed around as needed by library users and serves as an entry point to the library's functionality. This pattern is a good one to follow because it allows users to create multiple instances of your library and makes your library easier to test.

A perfect library may have no state, but in the real world, we often need ways to manage the library's internal state. In these cases, we should provide a way for users to manage that state. Also, we should make it clear how the state is managed and what the implications are for the user. If the state needs to be persisted or stored, we should provide a method for serializing and deserializing that state.

Examples of state we may need to handle include configuration, connection pools, caches, counters, and accumulators. Our library would likely have an entry point that accepts a context object, which would be passed around to the various functions in the library. Creating the context object would be handled by a factory function or a builder pattern, and the context object would be responsible for managing the state of the library. Consider the following example:

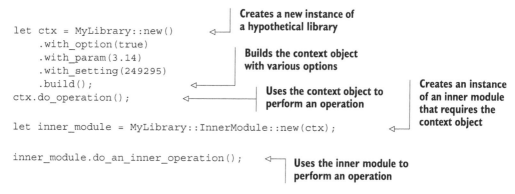

This example is a simple pattern for managing state in a library. The user creates the context object with a builder interface and is responsible for handling the context object and passing it around to the various functions in the library. The library doesn't need to leak the details of the context object, and the user can create multiple instances of the library with different configurations if necessary. If another module in the library requires access to the context object, the user can pass the context object to that module, such as an inner module in the preceding example.

6.10 *Consider the aesthetics*

First impressions matter, and the aesthetics of your library have a big effect on how people perceive it. Aesthetics aren't just about how the library looks but also about how it feels to use. A library that is easy to use, easy to understand, and debug will be perceived as more aesthetically pleasing than one that is difficult to use, understand, and debug.

The aesthetics of your library are influenced by many factors, including the naming of types, functions, and variables; the structure of the code; the documentation; the examples; and the overall design of the library. A library that is well organized, well documented, and easy to use is more aesthetically pleasing than one that is disorganized, poorly documented, and difficult to use.

When you're designing your library, consider the aesthetics of the code, the documentation, and the examples. Use consistent naming conventions; organize your code logically; and provide clear, concise, grammatically correct, error-free documentation. Documentation tools that produce good-looking documentation make this task a lot easier. Write examples that demonstrate how to use the library simply and straightforwardly. Consider the user experience of using your library, and strive to make it as pleasant as possible.

6.11 *Examining Rust library ergonomics*

Let's tie together some of what you've learned in the book so far by creating a library based on a previous code sample. This exercise is great for writing libraries. You can learn a lot simply by documenting and testing your code from the perspective of end users of your library. Also, I believe that forcing yourself to pay attention to the details from the perspective of other users enables you to produce better code. Writing libraries forces you to encapsulate, separate concerns, and create good interfaces.

You may be disappointed if you arrived here hoping to find a comprehensive list of all the dos and don'ts of creating libraries. I can't provide that list, but I can give you the skills you need to produce high-quality code.

6.11.1 *Revisiting linked lists*

We'll use the linked-list example from chapter 3 to form the basis of a library. We'll start by creating a library with `cargo new --lib linkedlist`. We'll copy the code from chapter 3 into `src/lib.rs`. Next, we'll create an integration test in our library. We'll create `tests/integration_test.rs` and populate it with the code from our previous test:

```
#[test]
fn test_linkedlist() {
    use linkedlist::LinkedList;
```

```
        let mut linked_list = LinkedList::new("first item");        ◄─┐  Error here:
                                                                       │  LinkedList is
        // ... snip ...                                                │  private.
}
```

We're using an integration test as opposed to a unit test because we want to test our library from outside the scope of the crate. The test code (which we copied directly from the old code) doesn't compile in its current state because we never considered visibility. The compiler reports the following error:

```
error[E0603]: struct `LinkedList` is private
  --> tests/integration_test.rs:3:21
   |
3  |     use linkedlist::LinkedList;
   |                     ^^^^^^^^^^ private struct
   |
note: the struct `LinkedList` is defined here
  --> /Users/brenden/dev/idiomatic-rust-book/c06/
   linkedlist/src/lib.rs:22:1
   |
22 | struct LinkedList<T> {
   | ^^^^^^^^^^^^^^^^^^^^^
```

This error message makes sense. Let's fix the visibility by adding `pub` to each method from the `impl<T> LinkedList<T> { … }` block and to the `LinkedList` struct itself. Keep in mind that the individual fields within the struct are still private, as everything is private by default in Rust. If we try to compile again, we get more errors. The first error looks like this:

```
error[E0446]: private type `Iter<'_, T>` in public interface
  --> src/lib.rs:41:5
   |
41 |     pub fn iter(&self) -> Iter<T> {
   |     ^^^^^^^^^^^^^^^^^^^^^^^^^^^^^^^ can't leak private type
...
62 | struct Iter<'a, T> {
   | ------------------ `Iter<'_, T>` declared as private
```

Ah, yes—we forgot to make the iterators public. Let's make the iterators public by adding `pub` to the three iterator structs: `Iter`, `IterMut`, and `IntoIter`.

After we make these changes, we've compiled our code successfully, and the test code will work. We had to fix the visibility to make our code into a proper library.

6.11.2 *Using rustdoc to improve our API design*

Next, we'll examine the API of our library. The best way is to generate documentation using `rustdoc`. Anyone who uses our library is likely to spend a lot of time looking at our docs, so it's essential to have good-quality documentation if we want anyone to be successful using our library.

We'll generate docs by running the `cargo doc` command, which places the generated HTML files in `target/doc` within the crate. We can open the file at `linkedlist/index.html`, which is the main landing page for our crate's documentation. We haven't written any documentation yet, so all we see is a blank page that lists the structs we marked with `pub` (figure 6.1).

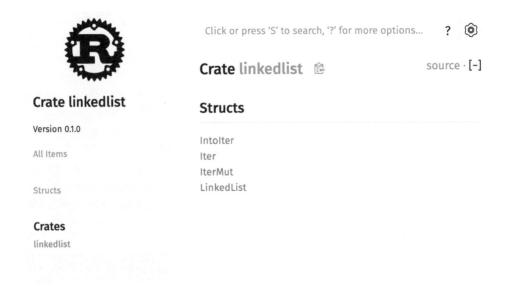

Figure 6.1 Empty documentation for our `linkedlist` crate

If we click the link for the `LinkedList` struct, we see the page shown in figure 6.2.

It's worth noting that even without doing anything to document our code, we have a fairly useful set of documentation. Simply because we listed our structures and methods (provided that we named them appropriately), a user can infer a lot about what our library does and how it works. This is especially true if we chose good names for our objects, methods, and traits. But we should write additional documentation anyway, no matter how self-explanatory we think our library is.

The first thing we should do is document the crate itself to tell anyone who looks at the documentation where to begin. We can document our crate by adding outer documentation to `lib.rs`. In Rust, outer documentation is provided by comments beginning with `//!`, and inner documentation is provided by comments beginning with `///`. Outer documentation applies to the outer scope of the file being documented, and inner documentation applies to the next item following the documentation comments.

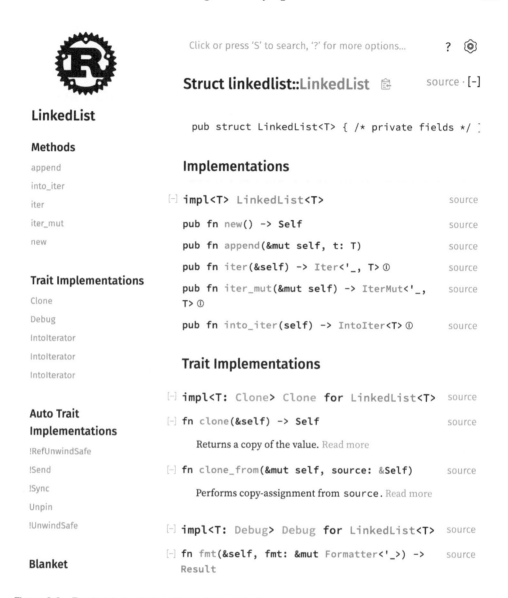

Figure 6.2   Empty `LinkedList` struct documentation

First, we'll add a top-level description of the crate and a high-level example of how to use the code. Let's update our code with the following at the top of `src/lib.rs`:

```
//! # linkedlist crate
//!
//! This crate provides a simple linked list implementation.
//!
//! The crate serves as a teaching example for the book [_Rust Advanced
//! Techniques_](https://www.manning.com/books/idiomatic-rust).
```

```
//!
//! ## Example usage
//!
//! ```rust
//! use linkedlist::LinkedList;
//!
//! let mut animals = LinkedList::new();
//! animals.append("chicken");
//! animals.append("ostrich");
//! animals.append("antelope");
//! animals.append("axolotl");
//! animals.append("okapi");
//! ```
```

After we regenerate the docs, our crate-level documentation looks like figure 6.3.

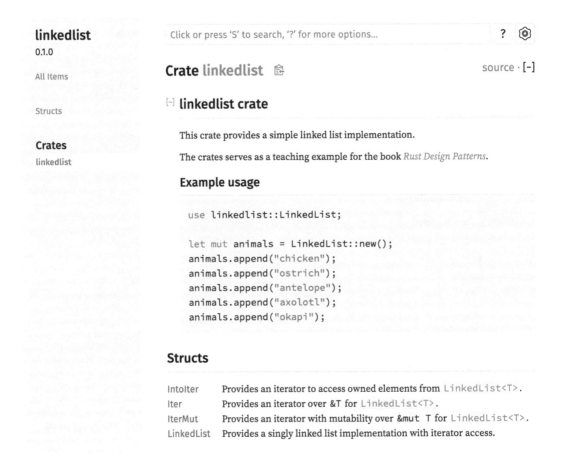

Figure 6.3 `linkedlist` **crate with top-level documentation**

Sweet! The documentation is starting to look like a real crate.

TIP   When you're working on documentation, use `cargo watch -x doc` to regenerate the docs automatically as you make changes. You can install `cargo-watch` with `cargo install cargo-watch` if you have not already.

Now that we have some documentation with a working example, we can test our docs. Every code sample in our docs is also an integration test. If we run `cargo test`, we see that our doc example automatically became a test (denoted by `Doc-tests linkedlist`):

```
$ cargo test
    Finished test [unoptimized + debuginfo] target(s) in 0.00s
     Running unittests src/lib.rs
     (target/debug/deps/linkedlist-2e0286b0918288ae)

running 0 tests

test result: ok. 0 passed; 0 failed; 0 ignored; 0 measured; 0 filtered out;
finished in 0.00s

     Running tests/integration_test.rs
     (target/debug/deps/integration_test-c95f81c9911957c8)

running 1 test
test test_linkedlist ... ok

test result: ok. 1 passed; 0 failed; 0 ignored; 0 measured; 0 filtered out;
finished in 0.00s

   Doc-tests linkedlist

running 1 test
test src/lib.rs - (line 10) ... ok

test result: ok. 1 passed; 0 failed; 0 ignored; 0 measured; 0 filtered out;
finished in 0.26s
```

Note that in our doc examples, we don't need to write a `main()` function. A small amount of preprocessing is applied by `rustdoc`, which wraps the code in `fn main() { … }` and creates the test code on the fly for execution with `cargo test`.

Let's talk about our API. When we wrote this code, we didn't think too much about how people might use it. One thing already stands out: our `new()` method on `LinkedList` looks a bit out of place. Why does `new()` take any parameters? I think we should emulate the behavior of other collections, like `Vec` in Rust. If we look at the documentation for `Vec::new()`, it states the following:

Constructs a new, empty `Vec<T>`.

The vector will not allocate until elements are pushed onto it.

For consistency, we should use the same pattern as `Vec`. Let's update our code by changing `new()` so that it returns an empty `LinkedList`. While we're at it, we should document our `LinkedList` as shown in the following listing.

Listing 6.1 `LinkedList` **with documentation**

```
/// Provides a singly linked list implementation with iterator access.
pub struct LinkedList<T> {
    head: Option<ListItemPtr<T>>,
}

impl<T> LinkedList<T> {
    /// Constructs a new, empty [`LinkedList<T>`].
    pub fn new() -> Self {
        Self { head: None }
    }
    /// Appends an element to the end of the list. If the list is empty,
    /// the element becomes the first element of the list.
    pub fn append(&mut self, t: T) {
        match &self.head {
            Some(head) => {
                let mut next = head.clone();
                while next.as_ref().borrow().next.is_some() {
                    let n = next.as_ref().borrow()
                        .next.as_ref().unwrap().clone();
                    next = n;
                }
                next.as_ref().borrow_mut().next =
                    Some(Rc::new(RefCell::new(ListItem::new(t))));
            }
            None => {
                self.head = Some(Rc::new(RefCell::new(ListItem::new(t))));
            }
        }
    }
    /// Returns an iterator over the list.
    pub fn iter(&self) -> Iter<T> {
        Iter {
            next: self.head.clone(),
            data: None,
            phantom: PhantomData,
        }
    }
    /// Returns an iterator over the list that allows mutation.
    pub fn iter_mut(&mut self) -> IterMut<T> {
        IterMut {
            next: self.head.clone(),
            data: None,
            phantom: PhantomData,
        }
    }
    /// Consumes this list returning an iterator over its values.
    pub fn into_iter(self) -> IntoIter<T> {
        IntoIter {
            next: self.head.clone(),
        }
    }
}
```

Notice that we also updated the `LinkedList` struct so that `head` is optional. We needed to make this change so that we'd have an empty instance because the preceding version assumed that we always had a `head` element. Figure 6.4 shows the updated documentation for `LinkedList`.

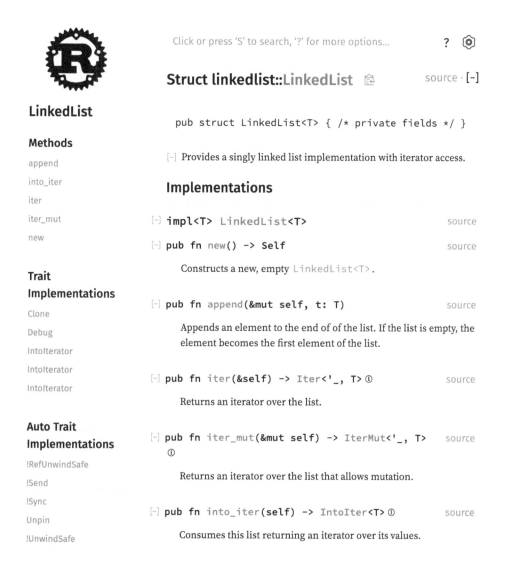

Figure 6.4   Documented `LinkedList`

That documentation looks good. We might want to consider adding many more features to our collection type, but let's focus on the most critical things. Two missing features come to mind: printing the contents of our list and cloning the list. Neither feature is as simple as it appears on the surface. We could use #[derive(Clone, Debug)], which would do an okay job of solving these problems, but it's not ideal. Let's talk about the problems separately.

If we want to implement Clone for our linked list, we have to consider what cloning a linked list means. Most likely, when someone calls clone() on the list, they intend to clone the structure and contents of the list, not the structure alone. In other words, we don't want to copy only the pointers to a new structure because they would still point to the same data.

To fix Clone, we have a couple of options: rewrite LinkedList so that it doesn't use Rc<RefCell<T>> or provide our own implementation for Clone instead of using #[derive(Clone)]. We want to continue using Rc<RefCell<T>> because it will make life easier if we decide to add more features to our list, so let's implement Clone ourselves. The definition for the Clone trait is as follows:

```
pub trait Clone {
    fn clone(&self) -> Self;
    fn clone_from(&mut self, source: &Self) { ... }
}
```

Neat. If we look at the Clone documentation a little more closely, we find the following statement about the clone_from() method:

> a.clone_from(&b) *is equivalent to* a = b.clone() *in functionality, but can be overridden to reuse the resources of* a *to avoid unnecessary allocations.*

This statement is good to know because I think it's easier to implement clone_from() than clone(). We can call clone_from() from our clone() implementation:

Note the trait bound on T. We provide Clone only for types that also implement Clone.

We clone the elements from the old list, self, into the new list by calling clone_from(), which we define below.

```
impl<T: Clone> Clone for LinkedList<T> {
    fn clone(&self) -> Self {
        let mut cloned = Self::new();
        cloned.clone_from(self);
        cloned
    }
    fn clone_from(&mut self, source: &Self) {
        self.head = None;
        source.iter().for_each(|item| {
            self.append(item.clone())
        });
    }
}
```

Creates the new list

The final expression returns the new list.

Setting the head of the list to None effectively resets the list.

We use our iterator to clone each value in the list and append the values to the target list, which is self.

That code makes things simple, and as a bonus, it follows the DRY (Don't Repeat Yourself) principle, so any changes to `clone_from()` are reflected by `clone()`.

6.11.3 *Improving our linked list with more tests*

We've added a bunch of new features, so we should test our code. Let's update our integration tests to test each feature separately. We'll start with the following listing, which tests the `iter()` method of our `LinkedList`.

> **Listing 6.2   Testing `iter()` for our `LinkedList`**

```
#[test]
fn test_linkedlist_iter() {
    use linkedlist::LinkedList;
    let test_data =
        vec!["first", "second", "third", "fourth", "fifth and last"];

    let mut linked_list = LinkedList::new();
    test_data
        .iter()
        .for_each(|s| linked_list.append(s.to_string()));

    assert_eq!(
        test_data,
        linked_list
            .iter()
            .map(|s| s.as_str())
            .collect::<Vec<&str>>()
    );
}
```

> We append a String to our test list even though we have Vec<&str>.

> We're using assert_eq!(), so the types we're comparing must match. Rather than convert our Vec<&str> to Vec<String>, we get a temporary Vec<&str> out of our linked list using collect().

Next, the following listing tests the mutable iterator method, `iter_mut()`, from our `LinkedList`.

> **Listing 6.3   Testing `iter_mut()` for our `LinkedList`**

```
#[test]
fn test_linkedlist_iter_mut() {
    use linkedlist::LinkedList;
    let test_data =
        vec!["first", "second", "third", "fourth", "fifth and last"];

    let mut linked_list = LinkedList::new();
    test_data
        .iter()
        .for_each(|s| linked_list.append(s.to_string()));

    assert_eq!(
        test_data,
```

```
        linked_list
            .iter_mut()
            .map(|s| s.as_str())
            .collect::<Vec<&str>>()
    );
}
```

The following listing tests the `into_iter()` method of our `LinkedList`.

Listing 6.4   Testing `into_iter()` for our `LinkedList`

```
#[test]
fn test_linkedlist_into_iter() {
    use linkedlist::LinkedList;
    let test_data =
        vec!["first", "second", "third", "fourth", "fifth and last"];

    let mut linked_list = LinkedList::new();
    test_data
        .iter()
        .for_each(|s| linked_list.append(s.to_string()));

    assert_eq!(
        test_data
            .iter()
            .map(|s| s.to_string())
            .collect::<Vec<String>>(),
        linked_list.into_iter().collect::<Vec<String>>()
    );
}
```

> For the into_iter() test, we'll convert the test data to Vec<String> instead of the other way around.

The following listing tests our implementation of the `Clone` trait.

Listing 6.5   Testing `Clone` for our `LinkedList`

```
#[test]
fn test_linkedlist_cloned() {
    use linkedlist::LinkedList;
    let test_data =
        vec!["first", "second", "third", "fourth", "fifth and last"];

    let mut linked_list = LinkedList::new();
    test_data
        .iter()
        .for_each(|s| linked_list.append(s.to_string()));

    let cloned_list = linked_list.clone();

    linked_list
        .into_iter()
        .zip(cloned_list.into_iter())
```

> To test whether our clone worked as intended, we use into_iter() because it returns the underlying owned value, which is what we want to check.

```
        .for_each(|(left, right)| {
            assert_eq!(left, right);
            assert!(!std::ptr::eq(&left, &right));
        });
}
```

Checks the values of
our original and cloned
lists to make sure they
match

Checks whether the original and cloned values are
different memory locations. This check is somewhat
redundant because it's not possible to have two
variables pointing to the same owned objects
in scope, but we'll make it anyway.

Our tests are passing, so we can move on.

6.11.4   *Making our library easier for others to debug*

Now let's talk about the Debug trait. Just for fun, let's see what happens if we try to use
#[derive(Debug)] and print our list with test data using dbg!(linked_list). The
output would look something like this:

```
[tests/integration_test.rs:20] linked_list = LinkedList {
    head: Some(
        RefCell {
            value: ListItem {
                data: RefCell {
                    value: "first",
                },
                next: Some(
                    RefCell {
                        value: ListItem {
                            data: RefCell {
                                value: "second",
                            },
                            next: Some(
                                .. snip ..
                            ),
                        },
                    },
                ),
            },
        },
    ),
}
```

Oh, my! That result isn't helpful at all. If someone is trying to use our linked list, this
output will make a big mess, especially if it's got deeply nested structures. We can't use
this code the way it is. Let's take a look at the Debug trait so we can go about implementing it:

```
pub trait Debug {
    fn fmt(&self, f: &mut Formatter<'_>) -> Result<(), Error>;
}
```

The interesting part of the Debug trait is Formatter. Rust gives us the Formatter tool, which takes care of the messy business of handling most of the formatting of our output. Formatter provides an easy way to format the debug output of lists with debug_list().

> **NOTE** You can find the complete reference for Formatter at https://mng.bz/67VG.

Let's implement the Debug trait by using Formatter::debug_list():

```
impl<T: Debug> Debug for LinkedList<T> {
    fn fmt(&self, fmt: &mut std::fmt::Formatter<'_>) -> std::fmt::Result {
        fmt.debug_list().entries(self.iter()).finish()
    }
}
```

With our new Debug implementation, the output from our tests using dbg!() looks like the following, which is a remarkable improvement:

```
[tests/integration_test.rs:20] linked_list = [
    "first",
    "second",
    "third",
    "fourth",
    "fifth and last",
]
```

Last, let's look at our documentation for the iterators we created. We didn't write any documentation for Iter, IterMut, or IntoIter. But if we look at the documentation that was generated, we see that the Iterator trait has provided a lot of functions for us, and those functions are already documented. Let's write a short description of each iterator for the sake of completeness (figure 6.5).

Now we've now got a decent-looking crate! Our library is quite Rustaceous, which we know because we've created good-quality documentation, provided implementations of key traits, and mirrored the API of Vec. Someone who's already familiar with Vec should be able to use our collection without much pain because we followed the existing patterns in the language.

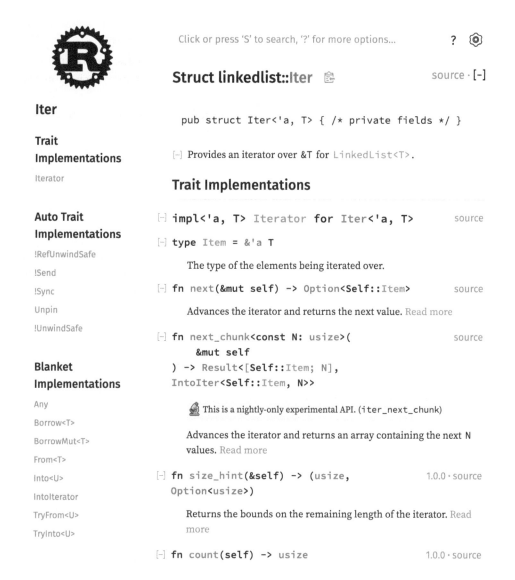

Figure 6.5 Documentation for our iterators

Summary

- Good library design is difficult, requires careful thought, and may require many iterations.
- Libraries should be designed with users in mind. We should strive to make our libraries easy to use and understand.
- Do one thing, and do it well. A library should have few responsibilities and should focus on solving a specific problem.

- Focusing on correctness is essential. We should use tools such as property-based testing and fuzzing to ensure that our libraries are correct.
- We should avoid excessive abstraction and stick to basic types whenever possible.
- Tools like Clippy and rustfmt can help us ensure that our code is idiomatic and easy to read.
- Examining popular crates is an excellent way to find inspiration for designing a library.
- Maintaining a library is an ongoing process, particularly if we want to publish a crate. We may need to make bug fixes, add features, or update our crate as new Rust features become available. Maintaining backward compatibility is essential, and we should follow semantic versioning to signal compatibility between versions.
- When designing libraries, we should pay special attention to how our APIs work from the perspective of our users. By providing good documentation with examples and comparing our own APIs with existing well-known APIs, we can create great libraries with surprisingly minimal work.

Part 3

Advanced patterns

At this point, I hope that you have taken time to experiment with the patterns in the preceding chapters and digest their concepts. The patterns in this part are more advanced and may require more practice and time to learn solidly. If I did my job right, you may have breezed through the book up to now. If not, don't worry; practice makes perfect.

You may also find that some of these patterns are more specialized and may not be as widely applicable. But it's still good to know and understand them, as you'll most likely encounter them in the wild. I think that the more time you spend working with Rust, the more value you'll get from these patterns.

Using traits, generics, and structs for specialized tasks

This chapter covers

- Using const generics
- Applying traits to external crate types
- Extending types with extension traits
- Implementing blanket traits
- Using marker traits to mark types with attributes
- Tagging with structs
- Providing access to internal data with reference objects

The previous chapters introduced several Rust advanced techniques. This chapter expands on some of those themes and explores more advanced design patterns. These patterns are useful in many circumstances, but you'll likely use them less frequently because they are more complex to implement and often apply to scenarios that you may not encounter often.

To use an analogy, the patterns discussed in the previous chapters describe standard tools you might find in every toolbox that can be used for a wide variety of jobs: hammer, pliers, screwdriver, power drill, and so on. The patterns discussed in this chapter are for more specialized jobs, such as the woodworking tools you would find in a wood shop: table saw, planer, lathe, band saw, and so on. Although these patterns are less useful for everyday Rust programming, it's good to explore them so that you'll know how to use them when you need them.

7.1   Const generics

Rust's *const generics* are a neat twist on generics that allow you to use constant values generically. Const generics solve a long-standing problem in languages with generics that occurs when you want to include a field in a structure that depends on a constant value (such as the length of an array). The constant value is known only at the time of instantiation, so without const generics, the only way to enable it would be to create a version of your struct for every desired size—which is exactly what many libraries do.

We can use const generics anywhere we have both a primitive constant and a generic parameter, such as to define the size of an array. We can use any size of integer-based primitive, such as `i32`, `u32`, and `usize`. We can also use `char` and `bool` types (which at the compiler level are equivalent to `u8` on most platforms), but floating-point values aren't permitted.

To understand const generics, let's explore the problem they solve. Suppose that we have a generic structure with an array of bytes, which we'll call a buffer:

```
struct Buffer {
    buf: [u8; 256],
}
```

Our buffer holds 256 bytes. What if we want to make it generic so that it can hold any type, not just bytes? Let's do that:

```
struct Buffer<T> {
    buf: [T; 256],
}
```

Presto! Easy enough. Now our buffer can hold 256 elements of anything. But wait a minute—what if we want the length of the array to be arbitrary? In other words, we should make the length of the array variable at the time of instantiation. One way would be to use a `Vec`, which can be resized at run time. The problems with using a `Vec` are that it requires heap allocation (whereas we can allocate a plain array on the stack) and it introduces a certain amount of overhead that we may not need, such as copying values instead of moving them.

If we know that the length of the array will never change throughout the life of our buffer (as is often the case), we can use a *const* generic parameter. Let's introduce a `LENGTH` parameter using const generics:

```
#[derive(Debug)]
struct Buffer<T, const LENGTH: usize> {
    buf: [T; LENGTH],
}
```

Now our struct has two generic parameters: the type of the array elements and the length. The LENGTH parameter can be treated like other generic parameters except that it's a constant value instead of a type, which results in some neat side effects. It creates a new distinct type when it's instantiated, for example, which is useful when we want to use Rust's type system with arbitrary-length arrays. We can provide concrete trait implementations for particular constant values, for example, which helps us avoid a whole class of programming errors, such as when a mismatch occurs in the provided and expected lengths of an array. We can specialize on particular constructions of this struct, such as with the From trait for an array of [u8; 256]:

```
impl From<[u8; 256]> for Buffer<u8, 256> {
    fn from(buf: [u8; 256]) -> Self {
        Buffer { buf }
    }
}
```

This implementation allows us to create a Buffer from an array of type [u8; 256] (but not any other type) by moving the array into the struct. Practically speaking, that approach isn't very useful. Instead, we probably want to implement a generic From and use specializations as needed:

```
impl<T, const LENGTH: usize> From<[T; LENGTH]> for Buffer<T, LENGTH> {
    fn from(buf: [T; LENGTH]) -> Self {
        Buffer { buf }
    }
}
```

This code lets us move an array of arbitrary type and length into our Buffer. This approach is quite useful, especially if we've built our code to work with Buffer, rather than raw arrays. We can test our buffer quickly with the following code:

```
let buf = Buffer::from([0, 1, 2, 3]);
dbg!(&buf);
```

◁── **Note that we don't need to specify the length parameter of 4; the compiler automatically infers it.**

Executing this code produces the following output:

```
[src/main.rs:14] &buf = Buffer {
    buf: [
        0,
        1,
        2,
        3,
    ],
}
```

Const generics make it easy to build custom array-based types with fixed lengths, which can save a lot of boilerplate.

7.2   *Implementing traits for external crate types*

When you start working with traits, you may get excited about them and start writing traits for all kinds of things. This experimentation can be fun until you run into a well-known design limitation of traits: you cannot implement a trait for types outside your crate.

This limitation exists for a good reason: if you could implement traits for any type, you could quickly wind up with multiple conflicting trait implementations for the same type. This situation could get worse as different crates evolve at different times, slowly diverging due to their conflicting implementations. The compiler could apply a heuristic to choose an implementation, but that approach would always be somewhat confusing and difficult to reason about; thus, the Rust language doesn't allow this as a matter of principle.

Worry not, however. Rust has a few features that let you implement equivalent behavior without creating conflicts, should you need to do so.

7.2.1   *Wrapper structs*

To unlock external type traits and continue to use the features of those types, we need to combine two different patterns: wrapper structs and the `Deref` trait. A *wrapper struct* is a struct that wraps another type. In its simplest form, it contains only one field of the thing being wrapped. After we create a wrapper struct, we can implement any trait we want for the wrapper.

We can use wrapper structs with `Deref` to implement traits for types from external crates to get around the limitation on external type traits and make an object behave like its subject. To demonstrate, let's wrap a `Vec`:

```
struct WrappedVec<T>(Vec<T>);          ⟵——— This construction of a struct is
                                             equivalent to a tuple.
```

> **NOTE**   A tuple struct, such as `WrappedVec`, is effectively equivalent to an ordinary tuple except that we've defined a new type with a name and can write `impl` blocks like any other struct.

That's easy enough. But if we try to use our `WrappedVec` like a `Vec`, it won't work:

```
let wrapped_vec = WrappedVec(vec![1, 2, 3]);                   Errors out on call to iter()
wrapped_vec.iter().for_each(|v| println!("{}", v));     ⟵——   with "method not found in
                                                               `WrappedVec<{integer}>`"
```

It makes sense that this code doesn't work: we haven't implemented `iter()`. We don't want to reimplement all the methods that `Vec` provides; we want to pass through to them from our wrapper struct.

7.2.2   *Using Deref to unwrap a wrapped struct*

The trick to making wrapper structs work nicely is implementing the `Deref` trait for our `WrappedVec`. When we implement `Deref`, the compiler automatically dereferences our wrapper when we call methods that don't exist. This approach is called `Deref` coercion, but you should take care to avoid overusing it. Implementing `Deref` is a piece of cake:

```
impl<T> Deref for WrappedVec<T> {
    type Target = Vec<T>;                 ◄─── The target type is what we want
    fn deref(&self) -> &Self::Target {         to dereference to automatically.
        &self.0                           ◄─┐
    }                                        │ The .0 on self denotes the first
}                                            │ element in the tuple struct. Each
                                             │ element in a tuple is unnamed.
```

Now we can call all the methods from `Vec`, such as `iter()`. Some limitations exist, however. For one, we can't use methods that take `self` by value, such as `into_iter()`. For that purpose, you'll need to implement the `into_iter()` method:

```
impl<T> WrappedVec<T> {
    fn into_iter(self) -> IntoIter<T> {
        self.0.into_iter()
    }
}
```

To call `Vec` methods that take `&mut self`, you need to implement the `DerefMut` trait, which is nearly the same as `Deref`. We can write a quick test for our wrapped vector:

```
let wrapped_vec = WrappedVec(vec![1, 2, 3]);             Our WrappedVec doesn't
wrapped_vec.iter().for_each(|v| println!("{}", v));      have any iterator methods,
wrapped_vec.into_iter().for_each(|v| println!("{}", v)); but we can call them from
                                                          Vec just like an ordinary
                                                          vector.
```

Running the preceding code produces the following output:

```
1
2
3
1
2
3
```

7.3 Extension traits

Extension traits are traits that add functionality to types and traits outside the crate in which they're defined. An example use of extension traits is adding features to standard library types, such as adding a method to the core type `Vec`. Extension traits typically follow a naming convention that uses the `Ext` postfix. You may encounter extension traits in crates that provide features for upstream crates or the standard library.

To illustrate an extension trait, we'll extend `Vec` by adding a new trait, `ReverseExt`, to which we'll add a `reversed()` method that returns a reversed copy of the vector. Our trait definition is as follows:

```
pub trait ReverseExt<T> {                  Our reversed() method
    fn reversed(&self) -> Vec<T>;     ◄─┐  returns a Vec<T>.
}
```

For simplicity, we return `Vec<T>` in this example. To improve this interface, you may want to add a second generic parameter for the returned container type, similar to how the `collect()` and `collect_into()` methods from Rust's `std::iter::Iterator` are implemented.

In practice, we might write a library that exports this trait with one or more implementations, which can be imported and used elsewhere. We don't necessarily need to write a library to use extension traits; we can also use them within our crate or application without exporting them. Let's implement `ReverseExt` for `Vec`:

```
impl<T> ReverseExt<T> for Vec<T>
where
    T: Clone,
{
    fn reversed(&self) -> Vec<T> {
        self.iter().rev().cloned().collect()
    }
}
```

We place a Clone trait bound on T so we can clone each item in the Vec.

To reverse the vector, we simply obtain an iterator, reverse it with rev(), clone each item, and collect the result in a new Vec.

We can test this code as follows:

```
let forward = vec![1, 2, 3];
let reversed = forward.reversed();
dbg!(&forward);
dbg!(&reversed);
```

When we execute the code, we get the following output, as expected:

```
[src/main.rs:17] &forward = [
    1,
    2,
    3,
]
[src/main.rs:18] &reversed = [
    3,
    2,
    1,
]
```

Another way to use extension traits is to apply them to another trait rather than a type. Following the preceding example, we can add a `to_reversed()` method to `std::iter::DoubleEndedIterator`:

```
pub trait DoubleEndedIteratorExt: DoubleEndedIterator {
    fn to_reversed<'a, T>(self) -> Vec<T>
    where
        T: 'a + Clone,
        Self: Sized + Iterator<Item = &'a T>;
}
```

We use a supertrait (which we'll discuss later) to limit the scope of our trait to apply only to DoubleEndedIterator.

The iterator item type and lifetime need to match T, and the iterator needs the Sized bound.

We need to require the Clone trait bound for T, the target type.

```
impl<I: DoubleEndedIterator> DoubleEndedIteratorExt for I {
    fn to_reversed<'a, T>(self) -> Vec<T>
    where
        T: 'a + Clone,
        Self: Sized + Iterator<Item = &'a T>,
    {
        self.rev().cloned().collect()
    }
}
```

← **Nearly identical to the previous version except without the call to iter()**

We can test this extension trait as follows:

```
let other_reversed = forward.iter().to_reversed();
dbg!(&other_reversed);
```

This code, when executed, produces the same expected output:

```
[src/main.rs:38] &other_reversed = [
    3,
    2,
    1,
]
```

One nice result of applying an extension trait to another trait (as opposed to a type) is that we can use this trait on any type that implements the `DoubleEndedIterator` trait, which includes `Vec`, slices, and `std::collections::LinkedList`, among others.

7.4 Blanket traits

Sometimes, we have especially generic traits in the sense that they apply to nearly any type, and for those traits, we may want to provide blanket implementations. A *blanket trait implementation*, unlike a concrete implementation, uses generic parameters. You can also have partial blanket implementations that specialize for some parameters but are generic for others.

We can use blanket traits to quickly and easily implement a trait for all types that satisfy our criteria. The criteria are specified in terms of trait bounds; our blanket trait implementation will apply to any type that implements the traits in our trait bound.

Some traits in the Rust standard library, for example, provide blanket implementations. Blanket trait implementations often depend on other traits or types, such as the `ToString` trait, which provides a blanket implementation as follows:

```
impl<T: Display> ToString for T {
    // ...
}
```

This implementation, lifted from the Rust standard library, depends on `Display`'s being implemented for `T`. For any type that provides `Display`, `ToString` is provided automatically (that is, you can call the `to_string()` method).

Creating a blanket implementation is relatively simple. We simply need to use generic parameters for all or part of the target type. We can create a blanket trait for all types in our crate, if we want:

```
trait Blanket {}
impl<T> Blanket for T {}
```

Implements Blanket for
all types in the crate

This example isn't too useful in its current form, but the code is quite correct. Blanket implementations are useful when we apply them to specific types or bind them to another trait by using trait bounds. Sometimes, we want to use blanket traits as markers, as described in section 7.5. Another use of blanket traits is to combine several other traits into one.

Blanket traits can be useful for library authors who want to give users features without implementing every possible combination of types. Using the `Buffer` example from section 7.1, we may want to provide a blanket trait to convert from `Vec<T>` to a `Buffer`.

> **Listing 7.1   Blanket trait implementation with const generics**

```
impl<T: Default + Copy, const LENGTH: usize> From<Vec<T>>
    for Buffer<T, LENGTH>
{
    fn from(v: Vec<T>) -> Self {
        assert_eq!(LENGTH, v.len());
        let mut ret = Self {
            buf: [T::default(); LENGTH],
        };
        ret.buf.copy_from_slice(&v);
        ret
    }
}
```

The size of the Vec must match
the declared **LENGTH** parameter.

copy_from_slice() uses memcpy() under
the hood, and requires the source and
target to have the same length.

This code provides blanket implementation for the `From` trait for a `Buffer` of any type or length, provided that we have a `Vec`. It allows us to convert a `Vec` to a `Buffer` by using `into()` or `from()`. The code also combines `Default` and `Copy`, two other traits that are frequently provided, so we can be reasonably confident that they will be available for most types. We can test our blanket trait quickly as follows:

```
let group_of_seven = vec![
    "Canada",
    "France",
    "Germany",
    "Italy",
    "Japan",
    "United Kingdom",
    "United States",
    "European Union",
];
let g7_buf: Buffer<&str, 8> = Buffer::from(group_of_seven);
dbg!(&g7_buf);
```

We need to specify the target buffer length
of 8 because the compiler doesn't know the
length of the vector at compile time; the
vector is variable-length.

NOTE   If you astutely noticed eight items in the Group of Seven list, that number isn't a mistake. For reasons that go beyond the scope of this book, the European Union is not enumerated.

Running the preceding code will produce the following output:

```
[src/main.rs:34] &g7_buf = Buffer {
    buf: [
        "Canada",
        "France",
        "Germany",
        "Italy",
        "Japan",
        "United Kingdom",
        "United States",
        "European Union",
    ],
}
```

For library authors, blanket trait implementations improve the usability of a library. But we don't need to stress about providing the most generic implementation or every imaginable concrete implementation. Rather, we should focus on handling the most common cases, as we did by providing `From` for `Vec`.

7.5   Marker traits

When you get comfortable with traits, you might start noticing the use of marker traits in other Rust projects. *Marker traits* are abstract traits that mark or indicate features or attributes about a type in Rust without necessarily providing any behaviors. (Marker traits are often denoted by their absence of methods.) Marker traits don't have a specific use case; they can be useful in many contexts.

The difference between marker traits and regular traits is that marker traits don't necessarily provide behavior. The `Sync` and `Send` traits, for example, are marker traits, but neither `Sync` nor `Send` provides methods or functionality itself. `Sync` and `Send` are special cases because you can't even implement them without using `unsafe`; only the compiler can do so safely.

One form of a marker trait provides a blanket implementation that combines other traits. If we want a shorthand way to indicate that a particular type implements a given set of traits, for example, we can mark it accordingly. Consider the trait shown in the following listing.

Listing 7.2   Full-featured marker trait

```
#[derive(
    Clone, Copy, Debug, Default, Eq, Hash, Ord, PartialEq, PartialOrd,
)]
struct KitchenSink;                 ◄——————    An empty struct, for which we
                                               derive all the derivable traits
trait FullFeatured {}    ◄——┐                  from the standard library
                            │  An empty
                            │  marker trait
```

```
impl<T> FullFeatured for T where
    T: Clone
        + Copy
        + std::fmt::Debug
        + Default
        + Eq
        + std::hash::Hash
        + Ord
        + PartialEq
        + PartialOrd
{
}
```
◁── **A blanket implementation of our marker trait for any type that implements all the bounded traits**

This listing creates an empty marker trait called `FullFeatured`. Then we can create a blanket implementation for any time it meets the trait bounds, which is a list of all the derivable traits. Our `KitchenSink` unit struct is intentionally empty for this example, but we have derived every derivable trait (of the traits provided by the standard library) with the `#[derive(…)]` attribute for it. Now we can use our marker trait whenever we want to make sure that all those features are implemented without listing all of them every time:

```
#[derive(Debug)]
struct Container<T: FullFeatured> {
    t: T,
}
```
◁── **Specifies the FullFeatured trait bound for T**

This code creates a container type, which holds a single element. We've restricted the type of that element to types that provide the `FullFeatured` trait. We haven't explicitly implemented this trait; we're relying on our blanket implementation. We can test it as follows:

```
let container = Container { t: KitchenSink {} };
println!("{:?}", container);
```

Running the preceding code produces the following output:

```
Container { t: KitchenSink }
```

Marker traits don't have to be empty, though they often are. You can certainly treat traits that do have methods as marker traits, but conflating them may confuse other people. As a general rule, marker traits should be empty (contain no methods or types).

Supertraits

At this point it's worth discussing *supertraits*, which specify traits composed of other traits, as we did in the example with the `FullFeatured` trait.

We can use supertraits when we want to combine other traits into one supertrait. This approach can simplify code elsewhere, such as allowing us to reduce the number of distinct traits required for specifying trait bounds. Trait bounds can become quite complex, and we can use supertraits to consolidate a list of required traits.

To create a supertrait, we create a trait and specify a list of dependent traits, similar to trait bounds. A marker supertrait that combines `Clone` and `Debug` looks like this:

```
trait CloneAndDebug: Clone + Debug {}
```

The difference between using supertraits and providing blanket implementations with trait bounds (as we did with `FullFeatured`) is that supertraits give us slightly less flexibility (due to compiler strictness) and a little more convenience. With supertraits, we can't derive the `CloneAndDebug` trait unless our type implements both `Clone` and `Debug`. Using a blanket implementation instead allows us to make special exceptions for specific types. We can still derive our `FullFeatured` trait for any type, but the compiler won't enforce anything as it will with supertraits.

When choosing between supertraits and explicit implementations using trait bounds, as with `FullFeatured`, you should prefer supertraits if all you need is an alias for a set of existing traits. Also, supertraits allow us to provide default implementations for trait methods that use dependent traits.

We can update our `CloneAndDebug` trait to print a cloned copy of itself and return it:

```
trait CloneAndDebug: Clone + Debug {
    fn clone_and_dbg(&self) -> Self {
        let r = self.clone();
        dbg!(&r);
        r
    }
}
```

7.6 Struct tagging

Sometimes, we use structs to tag or mark generic types (those with generic parameters). This approach is called *struct tagging*. With struct tagging, we can use empty structs (also called *unit structs*) to tag a generic type by including the tag as an unused type parameter; the tag itself contains no state and may never be instantiated.

Like marker traits, the structs we use for tagging are typically empty; they're used to define state within the type system itself. The trick is that although we're using an abstraction intended to hold state (in this case, a struct), we're not holding any runtime state within the struct; instead, we're enabling the struct to be used as a generic type parameter.

As with marker traits, we're using one of Rust's core abstractions in a way that is somewhat perpendicular to its main purpose. By doing so, however, we can unlock some interesting programming patterns at compile time and in a type-safe manner. In C++ parlance, this approach is a form of *template metaprogramming*, such as that used by Boost's MPL (https://mng.bz/oevN).

We can use struct tagging when we want to perform compile-time computation without using macros. Struct tagging introduces a bit more complexity but has the advantage of being type-safe and checked by the compiler. If you're writing a library, you can build interfaces that are checked for correctness at compile time rather than run time, which can lead to more robust software. To illustrate the use of struct tagging, let's model a light bulb that has two states: on and off.

Listing 7.3   Modeling a light bulb with struct tagging

```
struct LightBulb<T> {                          A struct to model a light bulb, with
    phantom: PhantomData<T>,                   a type parameter for the bulb state
}

                          A unit tag struct to
                          represent an on light bulb
struct On;

struct Off;               A unit tag struct to
                          represent an off light bulb
```

We can construct an instance of our bulb with `let bulb = LightBulb<Off> { ... }`, which represents a light bulb in the off state. This kind of abstraction can be useful when we need to keep software state in sync with external state, such as when we're managing an external device (such as a light bulb) with software. Modeling with types rather than variables allows us to use the compiler to check that all our states and transitions are valid, as I'll explain in detail throughout the rest of this chapter.

Our code so far is okay, but we probably want to create a marker trait for the bulb state and add a trait bound. We should also give T a name that's more descriptive.

Listing 7.4   Adding a trait to our light-bulb model

```
trait BulbState {}                             We've added a marker
                                               trait for the bulb state.

struct LightBulb<State: BulbState> {           We've set a trait bound for State
    phantom: PhantomData<State>,               on our LightBulb to be a type
}                                              that provides the BulbState trait.

struct On {}
struct Off {}
                                               We'll implement the
impl BulbState for On {}                       BulbState marker trait for
impl BulbState for Off {}                      our on and off states.
```

This pattern will be extra useful if we start using the type state to create methods. Suppose that we want to transition the light bulb between on and off states. We can implement a state transition from on to off and vice versa.

Listing 7.5   Adding state transitions

```
impl LightBulb<On> {                                   We create a concrete
    fn turn_off(self) -> LightBulb<Off> {              implementation for a
        LightBulb::<Off>::default()                    lightbulb in an on state.
    }
}
              We define a turn_off() method that consumes
              this bulb and returns a new one in the off state.
```

```
    fn state(&self) -> &str {          ◁─┐  We've added a method to
        "on"                               return the name of this
    }                                       state for convenience.
}

impl LightBulb<Off> {                  ◁─┐  We define the same
    fn turn_on(self) -> LightBulb<On> {      methods for the inverse
        LightBulb::<On>::default()           state to switch from on
    }                                        to off.
    fn state(&self) -> &str {
        "off"
    }
}
```

Note that in this example, both the `turn_off()` and `turn_on()` methods take an
owned `self`, which consumes the `LightBulb` and returns a new one. We cannot
change a type parameter on generic structures, so we need to create and destroy them
instead. Last, we can test our new creation:

```
let lightbulb = LightBulb::<Off>::default();
println!("Bulb is {}", lightbulb.state());
let lightbulb = lightbulb.turn_on();
println!("Bulb is {}", lightbulb.state());
let lightbulb = lightbulb.turn_off();
println!("Bulb is {}", lightbulb.state());
```

Running this code produces the following output:

```
Bulb is off
Bulb is on
Bulb is off
```

Neat! The big advantage of using this pattern is that we gain the advantage of having
the type system check our states for us. We can use this pattern to build a type-safe
state machine, as discussed in chapter 8.

7.7 Reference objects

Reference objects provide a reference to interior data. We may want to use a reference
object to permit partial borrowing of interior data without providing public access. In
other words, we can wrap the private interior data in a public reference object to avoid
introducing a leaky abstraction or making the internal data public. Reference objects
typically use the `Ref` postfix in their name, which identifies them as holding references.

Figure 7.1 illustrates how reference objects maintain public and private data access
boundaries while providing a way to reference data (partially or entirely) within
an object.

We use reference objects to allow consumers of our API to share data via references without exposing the internal data structures or implementation details. Typically, these reference objects are accepted by our API in interfaces that operate on

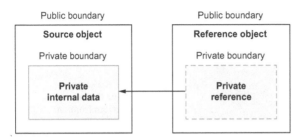

Figure 7.1 Reference objects

that data, so we can avoid making copies of data unnecessarily in certain circumstances. These reference objects are intended to be used only with the API from which they originate.

Suppose that we have two structs: `Student` and `StudentList`. Our `StudentList` is public and contains a `Vec`, but `Student` is private because we don't want to leak its data. The basic object definitions look like this:

```
#[derive(Debug)]
struct Student {
    name: String,
    id: u32,
}

#[derive(Debug)]
pub struct StudentList {
    students: Vec<Student>,
}
```

Now suppose that we want to design our code to obtain references to individual students within the list of students, but we don't want to provide direct access to internal data. We might have methods that operate on the reference objects and can perform operations, but the data can't be accessed directly. Let's create a public reference object as follows:

```
#[derive(Debug)]
pub struct StudentRef<'a> {
    student: &'a Student,
}
```

The lifetime parameter 'a lets us hold this reference for the lifetime of the Student object.

At this point, we have our basic reference object, `StudentRef`. We can test it as follows:

```
let student = Student {
    name: "Walter".into(),
    id: 582,
};
let student_ref = StudentRef { student: &student };
dbg!(&student);
dbg!(student_ref);
```

When we execute the code, we'll get the following output:

```
[src/main.rs:59] &student = Student {
    name: "Walter",
    id: 582,
}
[src/main.rs:60] student_ref = StudentRef {
    student: Student {
        name: "Walter",
        id: 582,
    },
}
```

This example works as expected, but we should make it a little more realistic. First, we'll add a constructor and accessors to the `Student` object.

Listing 7.6 Student with constructor and accessors

```
#[derive(Debug)]
struct Student {
    name: String,
    id: u32,
}

impl Student {
    fn new(name: String, id: u32) -> Self {
        Self { name, id }
    }

    fn name(&self) -> &str {
        self.name.as_ref()
    }

    fn id(&self) -> u32 {
        self.id
    }
}
```

Next, we need a way to obtain a reference from a `Student`. We'll create a `to_ref()` method.

Listing 7.7 Implementing `Student::to_ref()` to obtain a reference

```
impl<'a> Student {                          ◄─┐ Note the lifetime parameter 'a.
    fn to_ref(&'a self) -> StudentRef<'a> {   ◄─── The same 'a lifetime
        StudentRef::new(self)   ◄─┐                parameter is used for the
    }                            │                 method receiver self and
}                  We haven't created the          the returned StudentRef.
             StudentRef::new() method yet.
```

Next, we'll add a constructor that accepts a list of tuples, and we want to provide access to individual students from our `StudentList`. It would be convenient to look up students by ID or name, so let's add those methods.

Listing 7.8 `StudentList` with constructor and find methods

```
#[derive(Debug)]
pub struct StudentList {
    students: Vec<Student>,
}

impl StudentList {
    pub fn new(students: &[(&str, u32)]) -> Self {
        Self {
            students: students
                .iter()
                .map(|(name, id)| {
                    Student::new((*name).into(), *id)
                })
                .collect(),
        }
    }
}

impl<'a> StudentList {
    fn find<F: Fn(&&Student) -> bool>(
        &'a self,
        pred: F,
    ) -> Option<StudentRef<'a>> {
        self.students.iter()
            .find(pred)
            .map(Student::to_ref)
    }
    pub fn find_student_by_id(&'a self, id: u32) -> Option<StudentRef<'a>> {
        self.find(|s| s.id() == id)
    }
    pub fn find_student_by_name(
        &'a self,
        name: &str,
    ) -> Option<StudentRef<'a>> {
        self.find(|s| s.name() == name)
    }
}
```

We'll accept a slice of tuples to initialize the list.

Each tuple gets mapped to a new student.

Note the lifetime parameter 'a.

The lifetime parameter 'a needs to match for self and StudentRef.

Iterator::find() stops when the predicate returns true.

We map Some(student) to StudentRef using the Student::to_ref() method.

Both methods call the private find() method passing a closure, with nearly identical implementations differing only in the search parameter.

Note that `StudentList::find_student_by_id()` and `StudentList::find_student_by_name()` are nearly identical except for the `id` and `name` parameters, which we refactored into a private method, `find()`, that accepts a predicate closure. Let's test what we have so far with the following code:

```
let student_list =
    StudentList::new(&[("Lyle", 621), ("Anna", 286)]);

dbg!(&student_list);
dbg!(student_list.find_student_by_id(621));
dbg!(student_list.find_student_by_name("Anna"));
```

When we execute the code, we get the following output:

```
[src/main.rs:84] &student_list = StudentList {
    students: [
        Student {
            name: "Lyle",
            id: 621,
        },
        Student {
            name: "Anna",
            id: 286,
        },
    ],
}
[src/main.rs:85] student_list.find_student_by_id(621) = Some(
    StudentRef {
        student: Student {
            name: "Lyle",
            id: 621,
        },
    },
)
[src/main.rs:86] student_list.find_student_by_name("Anna") = Some(
    StudentRef {
        student: Student {
            name: "Anna",
            id: 286,
        },
    },
)
```

Everything looks good so far. Let's finish our `StudentRef` by adding a constructor.

Listing 7.9 `StudentRef` **with constructor**

```
#[derive(Debug)]
pub struct StudentRef<'a> {
    student: &'a Student,
}

impl<'a> StudentRef<'a> {
    fn new(student: &'a Student) -> Self {
        Self { student }
    }
}
```

Last, we can create a public function that operates on private data by using `Student-Ref`, without leaking the interior `Student` object to the caller. We could implement the `PartialEq` trait to check equality by student ID numbers, as follows:

```
impl<'a> PartialEq for StudentRef<'a> {
    fn eq(&self, other: &Self) -> bool {
        self.student.id() == other.student.id()
    }
}
```

We can test our `PartialEq` as follows:

```
let student_ref_621 = student_list.find_student_by_id(621).unwrap();
let student_ref_286 = student_list.find_student_by_id(286).unwrap();
dbg!(student_ref_286 == student_ref_621);
dbg!(student_ref_286 != student_ref_621);
```

Running this code produces the following output:

```
[src/main.rs:99] student_ref_286 == student_ref_621 = false
[src/main.rs:100] student_ref_286 != student_ref_621 = true
```

On a final note, it's possible to create mutable reference objects, but I'll leave that task to you as an exercise. Mutable reference objects are nearly the same except that they typically use the `MutRef` name postfix; you'll need to add the `mut` keyword to all references to satisfy the borrow checker (`&mut` and `&'a mut` as needed).

Summary

- Const generics allow us to use constant values as type parameters, unlocking features such as fixed-length arrays of arbitrary size.
- It's not possible to implement a trait for types outside our crate, but we can work around this limitation using wrapper structs and the `Deref` and `DerefMut` traits.
- Extension traits extend or alter the behavior of external types or traits, such as the standard library.
- We can implement a trait automatically for any combination of types by using generic implementations, known as blanket traits.
- Marker traits let us mark or denote types that have certain features or attributes, such as combining several other traits.
- We can use empty (or unit) structs to tag generic types by using the structs themselves as tags.
- Reference objects provide access to private interior data without transferring ownership or exposing internal private objects.

8

State machines, coroutines, macros, and preludes

This chapter covers

- Using traits to construct state machines
- Writing pausable functions with coroutines
- Implementing procedural macros
- Providing preludes to improve the usability of your crates

This chapter continues some of the themes from chapter 7 and builds on much of what we've learned in the book. We'll start by discussing state machines and coroutines. Then we'll introduce procedural macros, an advanced Rust feature that allows us to generate code at compile time. Last, we'll discuss preludes, which are a commonly used Rust library pattern to improve usability.

Rust's traits are powerful, and combined with generics, they let us build type-safe abstractions that allow us to guarantee correctness at compile time. This has some fairly significant implications, as we can avoid a host of problems that often plague software. State machines are robust ways to model stateful systems, and as we'll see in this chapter, it's surprisingly easy to build type-safe state machines in Rust.

State machines have always interested me, and I've used them many times, but I particularly like how easy it is to build a basic state machine in Rust without using additional crates or libraries. When building stateful systems in Rust, I create many small state machines as needed.

This chapter introduces Rust's coroutines, an upcoming experimental feature that's worth discussing because of its important future uses. Rust's coroutines may look familiar if you have previously encountered Python's generators.

8.1   *Trait state machine*

Now that we've explored traits and generics, we can start building some interesting abstractions on top of Rust's type system. One such abstraction, and arguably an incredibly useful one, involves building state machines. A *state machine* usually consists of a list of states and a set of transitions between states. We can define as many states or transitions as we want, but we can perform only valid transitions. Rust's type system enforces those rules.

Chapter 7 briefly demonstrated these rules with the light-bulb example; let's explore it further by modeling a user account session with a state machine, as shown in figure 8.1. We'll assume that we can have an anonymous or authenticated user.

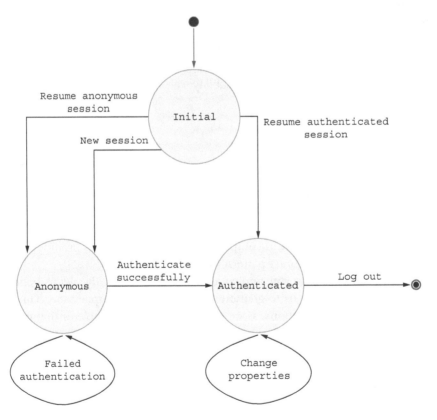

Figure 8.1   Modeling a user session with a state machine

Depending on which state the user is in, they may be able to perform various actions, such as changing account settings. Our session will include a session ID, which could map to user-side state (such as a cookie), and a session ID in a database, in addition to some arbitrary properties. Similar code could be used on either the client or server side. We'll create the structures shown in the following listing.

Listing 8.1   Modeling session state with traits and struct tagging

```
pub trait SessionState {}

#[derive(Debug, Default)]
pub struct Session<State: SessionState = Initial> {
    session_id: Uuid,
    props: HashMap<String, String>,
    phantom: PhantomData<State>,
}

#[derive(Debug, Default)]
pub struct Initial;
#[derive(Debug, Default)]
pub struct Anonymous;
#[derive(Debug, Default)]
pub struct Authenticated;
#[derive(Debug, Default)]
pub struct LoggedOut;

impl SessionState for Initial {}
impl SessionState for Anonymous {}
impl SessionState for Authenticated {}
impl SessionState for LoggedOut {}
```

> **We set the default session state to Initial.**

> **We'll keep a HashMap of arbitrary properties, which might be stored in a database.**

This listing defines four session states: `Initial`, `Anonymous`, `Authenticated`, and `LoggedOut`. Figure 8.1 shows the relationships between these states. We've added a `session_id` field to our `Session` struct, which will hold a universally unique identifier (UUID), provided by the `uuid` crate. Let's add some methods, beginning with the following listing.

Listing 8.2   Handling the initial state for `Session`

```
#[derive(Debug)]
pub enum ResumeResult {
    Invalid,
    Anonymous(Session<Anonymous>),
    Authenticated(Session<Authenticated>),
}

impl Session<Initial> {
    /// Returns a new session, defaulting to the anonymous state
    pub fn new() -> Session<Anonymous> {
        Session::<Anonymous> {
            session_id: Uuid::new_v4(),
            props: HashMap::new(),
```

> **An enum representing the result of the initial state transition**

> **These methods are limited to Session<Initial>, such as a session in the initial state.**

> **We provide a new() method for a new anonymous session.**

```
                phantom: PhantomData,
            }
    }
    /// Returns the result of resuming this session from an existing ID.
    pub fn resume_from(session_id: Uuid)
            -> ResumeResult {
        ResumeResult::Authenticated(
            Session::<Authenticated> {
                session_id,
                props: HashMap::new(),
                phantom: PhantomData,
            })
    }
}
```

Returns a ResumeResult to resume from an existing session

Here, we'd have to check the session_id against a database and return the result accordingly. For this example, we'll return a new authenticated session for testing purposes.

With this code, we can create a new anonymous session or resume from an existing authenticated one. In practice, the resume operation would involve a database lookup and validation for the session ID, but we'll omit those steps. Take a look at the code in the following listing.

Listing 8.3   Adding transitions for anonymous session

```
impl Session<Anonymous> {
    pub fn authenticate(
        self,
        username: &str,
        password: &str,
    ) -> Result<Session<Authenticated>,
        Session<Anonymous>> {
        // ...
        if !username.is_empty()
            && !password.is_empty() {
            Ok(Session::<Authenticated> {
                session_id: self.session_id,
                props: HashMap::new(),
                phantom: PhantomData,
            })
        } else {
            Err(self)
        }
    }
}
```

These methods are limited to instances of Session<Anonymous>.

We return a Result with either success or failure, and it consumes self.

Here, we would perform the authentication process, but we're simulating that process in this example. We use Session<Anonymous> as the error type, which allows us to indicate that authentication failed and the session is still in the anonymous state.

We simulate checking credentials by testing whether they're empty.

Last, examine the following listing.

Listing 8.4   Adding transitions for authenticated session

These methods are limited to instances of Session<Authenticated>.

```
impl Session<Authenticated> {
    pub fn update_property(&mut self,
                           key: &str,
                           value: &str) {
```

We can update properties for authenticated users, which might contain settings or preferences.

```
        if let Some(prop) = self.props.get_mut(key) {
            *prop = value.to_string();
        } else {
            self.props.insert(key.to_string(), value.to_string());
        }
        // ...
    }
    pub fn logout(self) -> Session<LoggedOut> {
        // ...
        Session {
            session_id: Uuid::nil(),
            props: HashMap::new(),
            phantom: PhantomData,
        }
    }
}
```

We would perform the actual property update here (such as writing to a database) and handle error/edge cases, but we're simulating the update in this example.

Calling logout() out will consume the session and return a logged-out session.

We would perform the logout process here, but we're simulating it in this example.

Now we have an excellent little state machine for handling sessions. We can run a quick test of our code as follows:

```
let session = Session::new();
println!("{:?}", session);
if let Ok(mut session) =
    session.authenticate("username", "password")
{
    session.update_property("key", "value");
    println!("{:?}", session);
    let session = session.logout();
    println!("{:?}", session);
}
```

If we run our test code, it prints something like the following:

```
Session { session_id: f0981fc3-3761-407f-b037-8759535acf87, props:
{}, phantom: PhantomData }
Session { session_id: f0981fc3-3761-407f-b037-8759535acf87, props:
{"some.preference.bool": "true"}, phantom: PhantomData }
Session { session_id: 00000000-0000-0000-0000-000000000000, props:
{}, phantom: PhantomData }
```

Sweet! This abstraction is fairly powerful, and we can build robust systems by modeling with state machines. State machines aren't panaceas, but they can make it much easier to reason about complex stateful systems. We can build a state machine combined with Rust's type system quickly and easily, with no need for additional libraries.

8.2   *Coroutines*

The upcoming coroutines feature in Rust provides pausable functions. With Rust's coroutines, we can create a closure that returns data to the caller through two separate paths: yielding and the function return path. We can also pause or terminate the coroutine immediately after yielding, which allows us to exit the coroutine early if necessary. Rust's coroutines will be familiar if you've used Python's generators. Coroutines are nightly-only and experimental, but they merit discussion due to their importance and potential utility.

> ### On the origins of coroutines
>
> Coroutines are loosely defined as functions that can pause and resume their execution. Coroutines are having a modern-day revival, but their origin can be traced back to Melvin Conway (of Conway's Law). Conway developed and coined the term *coroutine* in 1958. J. Erdwinn and J. Merner studied a similar idea at around the same time, but their paper "Bilateral Linkage," which described their work, was never published. In 1963, Conway more fully explained the concept of coroutines in his article "Design of a Separable Transition-Diagram Compiler," published in *Communications of the ACM*.
>
> The recent popularity of coroutines can likely be attributed to their use in Python's generator implementation (introduced in Python 2.5 in 2006) and Go's goroutines (2009), among others. Many other popular programming languages recently added similar coroutine implementations, including C++20, C# 2.0, Ruby Fibers, and PHP 5.5.
>
> Coroutines allow the introduction of concurrency without the need for threads, callbacks, or interprocess communication. They can be used to create complex control flows, such as cooperative multitasking and event loops.

Internally, coroutines are implemented by the Rust compiler, using a simple state machine. The overhead introduced by the compiler's implementation is minimal, consisting of a single enum for tracking the current coroutine's state.

NOTE For details on the current status of coroutines in Rust, refer to the Rust Unstable Book at https://mng.bz/ngnv.

There are many ways to use coroutines, but one application for them is to create iterators over data streams. Rust's coroutines are intended to enhance Rust's async/await features. They can also be used as building blocks for creating systems that use context switching or multiplexing, such as network programming and green threads. Rust's coroutine implementation is defined in the `std::ops::Coroutine` trait.

Listing 8.5 `std::ops::Coroutine` trait definition from Rust standard library

R is the closure's arguments, defaulting to unit ().
Self::Yield is the yield type, which will be unit () if unspecified.
Self::Return defines the closure return type.
The coroutine must be pinned.

```
pub trait Coroutine<R = ()> {
    type Yield;
    type Return;

    // Required method
    fn resume(
        self: Pin<&mut Self>,
        arg: R
    ) -> CoroutineState<Self::Yield, Self::Return>;
}
```

You don't need to implement the coroutine trait explicitly; the Rust compiler does that job for you when you create a closure containing a `yield` statement. For more complex scenarios, which we'll explore in listings 8.6 and 8.7, it's useful to understand how coroutines are implemented with the `Coroutine` trait.

Figure 8.2 illustrates the state machine for a coroutine, showing that when it's started by the first call to `resume()`, a coroutine can continue to yield values indefinitely until it returns, in which case it transitions to a completed state and no longer yields.

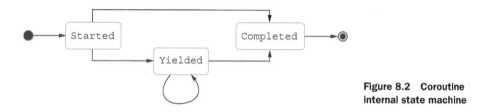

Figure 8.2   Coroutine internal state machine

A coroutine begins in the `Started` state and transitions to `Yielded` or `Completed` after the first call to `resume()`. If the coroutine yields a value, it transitions to the `Yielded` state and can continue to yield values indefinitely, as shown in figure 8.3. When the coroutine returns, it transitions to the `Completed` state and no longer yields. The coroutine can be resumed any number of times (such as in a loop), but when it has completed, it can't be resumed.

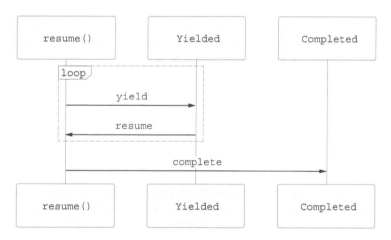

Figure 8.3   Coroutine sequence diagram

Let's look at the basic syntax for creating a coroutine, which is as simple as creating a closure with a `yield` statement and applying the `#[coroutine]` attribute to the closure. The following listing demonstrates a basic coroutine.

Listing 8.6 A basic coroutine in Rust

```rust
#![feature(coroutines,
           coroutine_trait,
           stmt_expr_attributes)]

use core::f64::consts::PI;
use std::ops::{Coroutine, CoroutineState};
use std::pin::Pin;

fn main() {
    let mut yield_pi = #[coroutine]
    || {
        yield PI;
        "Coroutine complete!"
    };
    loop {
        match Pin::new(&mut yield_pi).resume(()) {
            CoroutineState::Yielded(val) => {
                dbg!(&val);
            }
            CoroutineState::Complete(val) => {
                dbg!(&val);
                break;
            }
        }
    }
}
```

Coroutines are a nightly-only unstable feature that require a feature gate to enable them. We also need to enable expression attributes.

The #[coroutine] attribute must be applied to the closure.

A coroutine is defined by creating a closure.

The closure must have a yield statement. You can have multiple yields, but the types must match.

A coroutine also has a return type, which is distinct from the yield type. We can omit the explicit return, as the coroutine itself is a statement (returns the final expression).

Coroutines don't execute until they're resumed, and they must be pinned. Pinning prevents the coroutine from being moved in memory during execution.

When the coroutine returns from its closure, it has been completed.

A value can be yielded any number of times.

Our basic coroutine yields the number pi and then returns a string. It yields once on the first call to `resume()`, and on the second call, it returns and enters a completed state. Yielding a value is optional; we could also use the `yield` statement with no argument, which is equivalent to yielding `unit ()`. When we run the preceding code, we'll get the following output:

```
[src/main.rs:15:17] &val = 3.141592653589793
[src/main.rs:18:17] &val = "Coroutine complete!"
```

NOTE At the time of this writing, you need to implement an iterator on a coroutine yourself, but future versions of Rust may provide a blanket implementation of `Iterator` for the `Coroutine` trait.

To create a more interesting example of using coroutines, let's implement the `Iterator` trait on top of a coroutine, which allows us to use the `for` loop syntax in addition to all the other iterator features. To demonstrate, let's read the `Cargo.toml` file from the project we're working on. We'll define our `CargoTomlReader` object as shown in the following listing.

Listing 8.7   Implementing `Iterator` atop a coroutine

```
struct CargoTomlReader {
    coroutine:
        Pin<Box<dyn Coroutine<
            Yield = (usize, String),
            Return = ()
        >>>,
}

impl CargoTomlReader {
    fn new() -> io::Result<Self> {
        let file = File::open("Cargo.toml")?;
        let mut reader = BufReader::new(file);
        let mut line_number: usize = 0;

        let coroutine = Box::pin(
            #[coroutine]
            move || loop {
                let mut line = String::new();
                line_number += 1;
                match reader.read_line(&mut line) {
                    Ok(0) => return,
                    Ok(_) => yield (line_number, line),
                    _ => return,
                }
            }
        );
        let coroutine = Box::pin(
            #[coroutine]
            move || loop {
                let mut line = String::new();
                line_number += 1;
                match reader.read_line(&mut line) {
                    Ok(0) => return,
                    Ok(_) => yield (line_number, line),
                    _ => return,
                }
            },
        );
        Ok(Self { coroutine })
    }
}

impl Iterator for CargoTomlReader {
    type Item = (usize, String);
    fn next(&mut self) -> Option<Self::Item> {
        match self.coroutine.as_mut().resume(()) {
            CoroutineState::Yielded(val) => Some(val),
            CoroutineState::Complete(()) => None,
        }
    }
}
```

We use a trait object within a pinned box, specifying the yield and return type.

We'll keep track of the line numbers.

The closure must have the #[coroutine] attribute.

Our coroutine closure consists of a loop that will yield each line, and we move captured values into the closure.

Each pass through the loop increments the line number.

If BufReader::read_line returns 0, we've reached EOF, so we terminate with return.

If we get any value other than 0 from BufReader::read_line, we return the line and its number.

All other error cases result in the coroutine's completing.

We pass unit () to the coroutine as the starting argument.

We can test our `CargoTomlReader` with the following code to print each numbered line:

```
let cargo_reader = CargoTomlReader::new()?;
for (line_number, line) in cargo_reader {
    print!("{line_number}: {line}");
}
```

When we execute the code, we'll get the following output:

```
1: [package]
2: name = "coroutines"
3: version = "0.1.0"
4: edition = "2021"
5:
6: # See more keys and their definitions at https://doc.rust-lang.org/
cargo/reference/manifest.html
7:
8: [dependencies]
```

This example demonstrates a few key points about using coroutines with an iterator. Notably, because you need to pin the coroutine closure, using `Pin<Box<T>>` is a relatively easy way to handle this task. For any state we need inside the coroutine, we can initialize it at the beginning of the closure or use `move` to move captured variables into the closure, as I did in listing 8.7.

Coroutines are exciting and new but subject to change. Be careful when depending on this API, as it has not yet been stabilized. I can't speculate on when coroutines will be stabilized, but feel free to experiment with them and provide feedback to the Rust team, provided that you're willing to work with an unstable feature.

8.3 *Procedural macros*

Procedural macros are an advanced macro system in Rust that enables metaprogramming of arbitrary complexity, allowing all kinds of language extensions. We've used procedural macros quite a bit throughout the book, but we haven't discussed how they're implemented.

Many crates use procedural macros, and the most common use case (which we've seen many times in this book) is the `#[derive(...)]` attribute. Procedural macros are a big topic, warranting an entire book, so I'll simply touch on the basics here.

Creating a procedural macro involves writing a library that exports one or more macros and uses the `proc_macro` crate to implement the macro. The `proc_macro` crate is part of Rust, widely used throughout Rust and its ecosystem. You can't define a procedural macro in a binary crate; it must be in a separate library crate, though you may include it in your project as a workspace member. Procedural macros come in three forms:

- Function-like syntax, similar to declarative macros, such as `my_functionlike_macro!()`
- Derive macros, such as `#[derive(MyDerivableMacro)]`
- Attributes, such as `#[MyAttribute]`

Although there are no hard-and-fast rules about which form to use when, I'll break down the forms as follows:

- Function-like procedural macros—in the form `macro!()`, `macro!{}`, or `macro![]`—can be used anywhere in the code and are typically treated as functions or code blocks.
- Derive macros—in the form `#[derive(…)]`—can be used only with struct or enum declaration, but they allow the injection of arbitrary code following them.
- Attribute macros—in the form `#[MyAttribute]`—can be used to inject code just about anywhere, but they must be attached to an existing item. Attribute macros have one special feature: they allow you to supply arguments to the attribute.

Defining procedural macros requires providing Rust code that returns Rust syntax. That is to say, your macro definition is Rust code that writes Rust code. You must use the `proc_macro` crate to implement procedural macros, and your macros will be evaluated at compile time, as macros typically are.

Let's look at a simple procedural macro. We'll create a library with the following code:

```
use proc_macro::TokenStream;

#[proc_macro]
pub fn say_hello_world(_item: TokenStream)
        -> TokenStream {
    "println!(\"hello world\")".parse().unwrap()
}
```

This attribute indicates that the following function is a procedural macro.

parse() (which comes from FromStr) will parse this string into a TokenStream.

Procedural macro implementations are functions that take a TokenStream and return a TokenStream in its place.

This code operates on raw token streams. In practice, you wouldn't write a procedural macro this way; you'd use higher-level libraries, which we'll talk about in a moment. We also need to update `Cargo.toml` to indicate that this crate is a `proc_macro` crate. A `proc_macro` crate can export only a procedural macro, but you can include other crates as dependencies:

```
[lib]
proc-macro = true
```

Now we can test our code with

```
use hello_world::say_hello_world;
say_hello_world!();
```

When we run this code, it prints `"hello world"`. As I mentioned earlier, you probably wouldn't operate directly on `TokenStream`. Instead, two libraries are essential should you want to write procedural macros: `syn` and `quote`. The `syn` crate provides a parsing library to make it easier to work with source code, and the `quote` crate make generating Rust code a lot easier.

Let's examine a more realistic example of a procedural macro to demonstrate how all these pieces work together. In this example, we'll create our own derive macro, which will provide the name of the structure it's attached to. This macro is a form of reflection, and we're using a derive macro because it conveniently attaches to the declaration of a struct or enum.

First, we'll define a trait, which needs to be in a separate crate because a procedural macro library can't export anything other than procedural macros. The trait is shown in the following listing.

Listing 8.8 Trait to print the name of a struct

```
pub trait PrintName {
    fn name() -> &'static str;
    fn print_name() {
        println!("{}", Self::name());
    }
}
```

To implement our `PrintName` trait, we need to define the `name()` method, after which we can call `print_name()` to print the name of whatever it's implemented for. Next, let's write our macro.

Listing 8.9 Implementing the `PrintName` derive macro

Splits the generic clauses into their parts: impl
generics, type generics, and the where clause

Adds the necessary trait bounds only
if there are generic parameters

Converts the input token
stream to a syntax tree using
parse_macro_input!() which is
provided by the syn crate

```
#[proc_macro_derive(PrintName)]
pub fn print_name(input: TokenStream) -> TokenStream {
    let input = parse_macro_input!(input as DeriveInput);

    let generics = add_trait_bounds(input.generics);
    let (impl_generics, type_generics, where_clause) =
        generics.split_for_impl();

    let name = input.ident;

    let expanded = quote! {
        impl #impl_generics print_name::PrintName for #name #type_generics
                #where_clause {
            fn name() -> &'static str {
                stringify!(#name)
            }
        }
    };

    TokenStream::from(expanded)
}
```

We quote the actual trait implementation with all
the necessary parameters, including trait bounds.
quote!() is provided by the quote crate, and it
converts inline Rust syntax to a TokenTree.

We stringify the name of the type to
which we're applying the derive macro.
#name captures the name variable
value within the quoted block.

We convert the output of quote to a token
stream, which is provided by the quote crate.

```
fn add_trait_bounds(mut generics: Generics) -> Generics {
    for param in &mut generics.params {
        if let GenericParam::Type(ref mut type_param) = *param {
            type_param.bounds.push(
              parse_quote!(print_name::PrintName)
            );
        }
    }
    generics
}
```

The PrintName trait bound is added to all generic parameters for the target type.

In this listing, we include trait bounds for example purposes, but they're not required for the derive macro to work. Putting everything together, we can use a small integration test to verify that the code works:

```
use print_name::PrintName;
use print_name_derive::PrintName;

#[test]
fn test_derive() {
    #[derive(PrintName)]
    struct MyStruct;

    assert_eq!(MyStruct::name(), "MyStruct");
    MyStruct::print_name();
}
```

If we run `cargo expand --test test_derive` from the `print_name_derive` directory, we can examine the output of our macro:

```
fn test_derive() {
    struct MyStruct;
    impl print_name::PrintName for MyStruct {
        fn name() -> &'static str {
            "MyStruct"
        }
    }
    // ... snip ...
}
```

Nice! You can get much more elaborate with procedural macros, especially after you start handling attribute parameters or individual field attributes.

> **TIP**   In addition to checking the sample included with this book, consult the syn documentation at https://docs.rs/syn/latest/syn and the official Rust documentation at https://mng.bz/v8gx to learn more about implementing procedural macros. Also, Manning Publications has an excellent book called *Write Powerful Rust Macros*, by Sam Van Overmeire (https://www.manning.com/ books/write-powerful-rust-macros). For a real-life example of procedural macros in action, check out the `rocket` crate, which makes extensive use of procedural macros for its Rust web framework (https://crates.io/crates/rocket).

On a final note, procedural macros bring a lot of complexity. They are incredibly powerful, but that power is a double-edged sword. These macros can be tricky to debug when things go wrong, and they are unhygienic, which means that they can pollute or conflict with the namespace in which they're used. Because a procedural macro simply outputs code, which is injected before compilation, you must take care not to create conflicts or pollute the namespace.

8.4 Preludes

The last topic in this chapter is *preludes*—collections of useful types, functions, and macros provided for import into your code. When we're writing libraries, we can provide preludes to make it easy for people to get the most out of our library.

Some preludes, provided by the Rust language itself, are imported automatically, such as the standard library preludes. But I'm going to talk specifically about adding preludes to our crates rather than those from Rust.

> **TIP** For details on the Rust language preludes, consult the language reference at https://mng.bz/4JdB.

One reason we might use preludes when writing libraries is that it can be tricky to know which symbols to import. If we forget to import a trait, for example, we might find that our code doesn't compile or that functionality is missing, and figuring out what we missed can be frustrating. Preludes are implemented by means of re-exports, which is a way of exporting symbols from another module or crate.

Let's talk about use before we go deeper into implementing preludes. By now, we've already seen imports like this one:

```
use std::cell::RefCell;
use std::marker::PhantomData;
use std::rc::Rc;
```

It turns out that we can re-export anything imported with the use statement by adding the pub keyword, as we'd do with any other type or function:

```
pub use std::cell::RefCell;
pub use std::marker::PhantomData;
pub use std::rc::Rc;
```

When we re-export with a pub use ...; statement, the symbols imported by that use can be imported from outside that module, although we probably wouldn't want to re-export types from the standard library. It's also important to remember that if we want to import all the symbols exported by any module, we can use the wildcard (*) syntax with our imports:

```
use mylib::*;
```

This syntax imports everything exported from the top-level module of the `mylib` crate. Many libraries provide an explicit prelude module (usually named `prelude`) within their crates, and you would import from it as follows:

```
use mylib::prelude::*;
```

Using a separate prelude module is one way to avoid polluting the namespace. Let's walk through an example that implements the prelude trait in case you want to do that for your library. Suppose that you have a crate structured as follows:

```
$ tree
.
├── Cargo.lock
├── Cargo.toml
└── src
    ├── a.rs
    ├── b.rs
    └── lib.rs

1 directory, 5 files
```

This small library contains modules `a` and `b`. First, look at the following listing.

Listing 8.10   Listing for `lib.rs` without prelude

```
pub mod a;                    ◁──────┐   pub mod denotes that these
pub mod b;                           │   modules are available publicly
                                     │   (outside of the crate).
pub struct TopLevelStruct {}
```

Inside `a.rs` and `b.rs`, you created some empty public structs: `InnerA` and `InnerB`, respectively. From outside your crate, you can import the two structs from `a` and `b` with

```
use mylib::a::InnerA;
use mylib::b::InnerB;

// `InnerA` and `InnerB` are now within scope.
```

You haven't created your prelude yet. The prelude will be one module for the whole crate that exports all your most useful structs (in this case, `TopLevelStruct`, `InnerA`, and `InnerB`). You can make a module called `prelude`. The new crate structure looks like this:

```
$ tree
.
├── Cargo.lock
├── Cargo.toml
└── src
```

```
├── a.rs
├── b.rs
├── lib.rs
└── prelude.rs
```

1 directory, 6 files

Populate the `prelude` module with the code from the following listing.

Listing 8.11   Listing for `prelude.rs`

```
pub use crate::a::InnerA;
pub use crate::b::InnerB;
pub use crate::TopLevelStruct;
```

This code re-exports all the structs from your crate in one place. You also have to add `pub mod prelude;` to `lib.rs` to include `prelude.rs` in the crate. A user of this crate can import these three structs by using the wildcard `use` syntax:

```
use mylib::prelude::*;

// `InnerA`, `InnerB`, and `TopLevelStruct` are now in scope.
```

It's also possible to use aliases with `use` statements, as well as re-exports. You can re-export `TopLevelStruct` with a different name from your prelude if you want:

```
pub use crate::TopLevelStruct as AltStruct;
```

> **WARNING**   I would use this feature with caution. It's useful mostly if you want to use a different internal name versus an external name as a symbol.

You don't need to provide a separate prelude module to re-export symbols in this fashion; the pattern is simply used commonly in Rust. You can re-export symbols from any public module, including those outside your crate. Authors frequently export dependencies from third-party crates to make their crates easier to use.

Make no mistake—preludes are handy but can cause confusion because they hide some complexity at the expense of namespace pollution. Try not to abuse them.

If you're new to Rust and want to start publishing libraries, don't go crazy with preludes. With practice, you'll learn where they make the most sense. As a general rule of thumb, you shouldn't need them unless your crate provides lots of traits as part of its core functionality.

Summary

- Combining what we've learned about generics and traits, we can build abstractions such as state machines on top of Rust's type system.
- Coroutines are an experimental Rust feature, similar to Python's generators, that provides an alternative way to express pausable functions that can yield data.
- Procedural macros enable language extensions and metaprogramming well beyond what declarative macros can do.
- We can provide preludes for our libraries to make them a little more user-friendly by exporting the most useful parts of our library under one module.

Part 4

Problem avoidance

In the last part of the book, we focus less on which patterns to use and more on which patterns to avoid. Sometimes, it's worth sacrificing a little performance or memory use to build software that optimizes correctness, maintainability, and readability.

Fortunately, with Rust we don't have to sacrifice performance in most cases. Some people may argue that Rust has no real competitors in terms of safety and performance, so we're rarely sacrificing much when we dial down the speed a little in favor of correctness.

You may have found that the most challenging code to debug (and often the source of bugs) is code that is too clever for its own good. For this reason, this part focuses on avoiding patterns that are too clever, complex, or difficult to understand.

Immutability 9

This chapter covers

- Understanding the benefits of immutability
- Thinking in terms of immutable data and how it works in Rust
- Using traits to make nearly anything immutable
- Exploring crates that provide immutable data structures

Immutability is a powerful concept that can help everyone build better software. Immutability as it relates to writing software is the idea that after a value has been declared and assigned, it cannot be modified (or mutated). Contrast this concept with *mutability*, in which a value can be altered after it has been declared. In other words, values that can be changed are mutable, and values that are never changed are immutable.

Immutability is an important design pattern—and one of the most underloved and underappreciated ones at that. I feel that this pattern is so valuable, however, that I'm dedicating a chapter of this book to the subject, although it deserves an entire book. I won't be able to go into as much depth in this chapter as I'd like, but I'll leave you with a great starting point to explore the topic further.

In Rust, all declared variables are immutable by default, and you must explicitly opt in to mutability. For more complex data structures, however, you need to think a little harder about how you want to handle mutability and immutability. Some languages take immutability to the extreme by not allowing any mutations, but Rust (for better or worse) tends to leave the decision up to developers. Many programming languages and libraries have adopted immutability as a first-class feature, but Rust takes a more pragmatic approach, letting developers choose when and where to use it.

In this chapter, I'll discuss the benefits you derive from avoiding mutable data, look at some of the gotchas of trying to use data structures immutably, review Rust's approach to mutability and immutability, and show how to use Rust's features to make just about any ordinary data immutable. Finally, I'll describe some crates that provide immutable data structures, including some optimizations.

9.1   *The benefits of immutability*

If you haven't worked with languages or libraries that encourage immutability, the concept may seem a bit foreign. It's not uncommon for developers to be skeptical about immutability at first, but taking the time to understand the benefits is worthwhile. To help you make sense of immutability, I'll discuss the classes of problems that immutability can solve and how it solves them. Most software bugs fit into one or more of the following (nonexhaustive) classes:

- *Logic errors*—Mistakes, misunderstandings, or oversights in the code that lead to incorrect behavior. An example is the business logic in a program that calculates taxes on a purchase and mistakenly applies the wrong tax rate.
- *Race conditions*—Bugs that occur when shared data is not synchronized properly. This situation can lead to data corruption, deadlocks, and other problems, Race conditions are most often caused by concurrent access to shared mutable data, such as when multiple threads try to modify the same data at the same time or in the wrong order.
- *Unexpected side effects*—Unintended changes in the state of a program that occur when a function or method executes, leading to unexpected behavior and bugs that are difficult to track down. Side effects are often caused by functions that modify their arguments or global state or by functions that rely on global state. Any operations that involve I/O, such as reading from or writing to a file, are also side effects.
- *Memory safety problems*—Bugs that occur when a program tries to access memory that it shouldn't. Examples include programs that try to access memory that has already been freed, memory that they don't have permission to access, or memory outside the bounds of a data structure. These bugs can lead to crashes, data corruption, and (most concerning) security vulnerabilities.

Immutability can help solve all these problems in many cases. When it comes to logic errors, immutability helps by making it easier to reason about how the data in a program changes over time. When data is immutable, you can be sure that it won't change unexpectedly, which makes it easier to understand and predict the behavior of a program.

Race conditions occur only in programs that have shared mutable state. Immutability is an easy win here because if data is immutable, it can't be mutated while it's being shared. We may still need to have shared state, but if that state is never mutated, we don't have to worry so much about race conditions.

Side effects exist only when data is mutable. When we write a function that has no side effects, we call it a *pure function*. Pure functions are easier to reason about, easier to test, and easier to reuse. When we write code without side effects, we usually refer to it as being *purely* functional.

> **NOTE** If you're unfamiliar with functional programming, this concept can be tricky to grasp at first, but like most things, it gets easier with practice. When you begin writing purely functional code, you'll wonder how you ever managed to write code without it.

Pure functions have a nice property: they're *referentially transparent*, which means that you can replace a call to a function with the result of that function, and the program will behave the same way. Another way of thinking about this concept is that for any given set of inputs to a pure function, the output is always the same. Referential transparency allows us to introduce optimizations and refactorings that would be impossible with impure functions, and of course, it's much easier to reason about (and test) code when you know that a function will always return the same result for the same inputs. For software that's devoid of any side effects, an entire program can become deterministic—a powerful property that provides many benefits in testing and debugging.

Finally, immutability can help prevent memory safety problems. Quite often, memory safety problems arise when unexpected mutations occur, such as edge cases. Even when we write tests to cover the cases we know about, it's difficult to write tests for cases we don't know about. Strategies such as property testing and fuzz testing can help, but they can't cover every possible edge case.

When you take all these problems together, you can easily see why immutability is especially useful in parallel or concurrent systems, which often have to deal with them simultaneously. Several programming languages, libraries, and frameworks have adopted immutability as a core principle; the most notable examples include Erlang, Elixir, Haskell, Clojure, and Elm. These languages have a reputation for being reliable and for producing code that is easy to reason about. They also tend to be popular in domains in which reliability is paramount, such as telecom, finance, health care, and aerospace.

Some popular libraries and frameworks encourage immutability, including Redux, a popular state-management library for JavaScript and TypeScript applications. Redux, which is based on the principles of functional programming and immutability, is known for being reliable and easy to reason about. React, a wildly popular library for building interfaces, also encourages immutability, with an emphasis on writing purely functional components in recent versions of the library. The popular Immutable.js and Lodash libraries for JavaScript and TypeScript also provide utilities for working with immutable data.

These examples have influenced the design of the Rust programming language as well as some popular Rust libraries. If you've encountered these languages before, you may begin to see some similarities between them and Rust.

9.2   *Why immutability is not a magic bullet*

Immutability isn't free; it comes at a cost. The most obvious cost is that we often need to duplicate data when we want to change it, which can be expensive in terms of memory and CPU time and can make our code more complex. Immutability also has a higher cost in the sense that we need to spend more time up front thinking about how to structure our data and programs.

Rust generally attempts to minimize the cost of immutability by giving developers the option to opt in to mutability when they need it. Rust, however, does not enforce any patterns of immutability as part of its standard library. Rust's core data structures (such as Vec) are similar to vectors in C++ or C arrays, where mutability is in some ways encouraged and expected while using the data structure. This approach is different from that of languages that don't allow mutability, such as Erlang, Elixir, Haskell, Clojure, and Elm.

Some languages, such as Scala, provide a middle ground by allowing you to choose either immutable or mutable data structures, though Rust provides mutable structures only by default. Rust made this tradeoff (intentionally or not) within its standard library to provide a more familiar experience for developers coming from languages such as C and C++ and to provide a more efficient experience for developers who need to write high-performance code.

9.3   *How to think about immutable data*

Immutability as a high-level concept is somewhat incompatible with the way most of us think about data and, specifically, how computers handle data. Nearly all computers, big and small, from your desktop or laptop to your pocket computer (smartphone) to the largest supercomputers, are designed to handle mutable data and are based on the von Neumann architecture, shown in figure 9.1.

von Neumann architecture

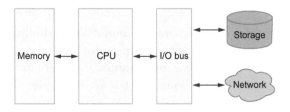

Figure 9.1   The von Neumann architecture

The von Neumann architecture is the basis of nearly all modern computers. It consists of a CPU that stores data and program instructions in the same memory. One implication is that the memory in which data is stored for fast access is finite and fundamentally mutable, which may differ from other storage systems, such as disk or tape. Because of the inherent constraints on memory, we have little choice but to reuse the same memory for different purposes, which is why we have mutable data.

If we had a hypothetical system that was append-only and could grow infinitely (such as an infinite tape, similar to a Turing machine), we could build an entirely immutable system. We could imagine a theoretical computer that uses an infinite tape as its memory, to which we can write only once. In this model, the tape is read from and written to but never modified—which, of course, is not how computers work in the real world.

Thus, immutability in the context of programming is merely an abstraction that exists only at a relatively high level. It's not built into the hardware architecture or the operating system. It's a concept that's enforced by the programming language and the libraries and frameworks that we use.

In the case of Rust, the borrow checker helps us keep track of which parts of our program are mutable and which parts are immutable, but for large or complex programs, it's still up to us to decide how we want to handle our data. This is true even of languages that are immutable by design because for any program to be useful, it must interact with the outside world, and the outside world is mutable, finite, and stateful.

The key to thinking about immutable data is to think about it in terms of ownership, borrowing, and the lengths you want to go to to enforce immutability. I think that most people will find the right balance by spending time working with the language and experimenting with different patterns and approaches.

It's also important to remember that immutability is not an all-or-nothing proposition. Some parts of your program can be immutable, and some can be mutable; some parts can be purely functional, and some parts are not. You may want to be strict for more critical components and less strict when you care more about performance than correctness.

It's up to you to decide what's best for your program, and having good judgment is a matter of having experience. This statement may be unsatisfying, but it's the reality of developing any skill, such as playing a musical instrument or working with computers and software; we all have to come to terms with it.

9.4   Understanding immutability in Rust

In Rust, all declared variables are immutable by default. The only exception occurs when the `unsafe` keyword is used to bypass the language guarantees. If you want to make a variable mutable, you must use the `mut` keyword to declare it as such. This feature cascades in such a way that data stored within an immutable structure is also immutable—a feature known as *inherited mutability*.

An exception to Rust's mutability rule applies when we use Rust's shareable mutable containers `Cell`, `RefCell`, and `OnceCell`. These containers enable *interior mutability*, which allows you to mutate the data inside an immutable container. This feature has some important implications and enables some powerful patterns, but understanding the tradeoffs is important. The problem with shareable mutable containers is that they allow hidden mutability in some circumstances, which may not always be desirable but isn't a problem in most situations.

The only practical difference between inherited and interior mutability is that the former is enforced by the compiler, and the latter is enforced at run time. As a general rule, you can consider values inside a `Cell`, `RefCell`, or `OnceCell` to be *optionally mutable*, even if the container itself is immutable. In other words, these containers provide the option to mutate the data inside them but don't require mutation.

`Cell` is slightly less devious than `RefCell` and `OnceCell` because it allows mutability only by way of replacement. You can't mutate the value inside a `Cell` directly, but you can replace the value with a new one. This distinction is subtle but important, and it's one of the reasons why `Cell` is considered to be a somewhat safer way to achieve interior mutability. `RefCell` and `OnceCell` allow you to obtain a mutable reference to the data they hold.

Rust's standard library doesn't provide data structures that are designed for immutability. Instead, Rust's data structures are generally what you'd expect of traditional mutable data structures. Both `Vec` and `HashMap` provide a variety of methods for mutable access, and they don't offer much in terms of features for working with data immutably aside from implementing the `Clone` trait. This situation may change in the future, but for now we have to work with what we have. Given Rust's focus on performance, it's often more efficient to work with mutable data structures. The basic pattern for implementing immutability in a language such as Rust that doesn't offer immutability as a core feature involves two steps:

- After a value is declared and assigned, it should not be modified in place.
- If you want to modify a value, you copy it and then modify the copy (which, confusingly, is still a mutation, but it's a mutation of a new value, not the original value).

For languages that provide immutability as a core feature, the language itself abstracts this process of copying and modifying, but fundamentally, the operation is still the same. We can use abstractions to hide the details of this process, and we'll explore some of these abstractions in sections 9.6, 9.7, and 9.8.

9.5   *Reviewing the basics of immutability in Rust*

Let's take a moment to review the basics of immutability in Rust, using some code samples. Although this process may seem trivial, it's good to review the basics now and then to understand why we do what we do (which some people call "first principles" thinking).

An immutable operation typically involves assigning the result of a computation to a new value. Suppose that we want to increment the value of the variable x immutably, which we could do by declaring a new variable y with `let y = x + 1`. A mutable operation would change the value of x directly with `x += 1`. In this example, it appears that `x += 1` is a shorter, more efficient way to increment x, but it's also more error-prone and can be more difficult to reason about in larger codebases or more complex scenarios.

Rust also allows you to shadow a variable, which is a way to declare a new variable and assign a value using the same name as an existing variable. This pattern is common in Rust, and it's often used to convert a mutable variable to an immutable one. You can shadow a mutable variable with an immutable, as the following listing demonstrates.

Listing 9.1 Basics of immutability in Rust

```
let x = 1;
dbg!(x);
let y = x + 1; // y = 2
dbg!(y);

// x += 1;                                    Uncommenting this
                                              line will result in a
// error: cannot assign twice to immutable variable `x`   compilation error.

let mut x = x; // x = 1        Shadows x with a
x += 1; // x = 2              mutable variable
dbg!(x);
```

Running this code produces the following output:

```
[src/main.rs:14] x = 1
[src/main.rs:16] y = 2
[src/main.rs:22] x = 2
```

Digging deeper, it's important to note that Rust's mutability semantics also apply across function calls, with one minor difference. Owned values can be switched from immutable to mutable when they are moved, but the caller of a function has no say in this process.

Listing 9.2 Mutability across function calls

```
fn mutability(                    Variable a is moved into the
                                  function and is immutable.
    a: i32,      // immutable
    mut b: i32,  // mutable       Variable b is moved into the
) {                               function and is mutable.
    // a += 1; // error: cannot assign twice to immutable variable `a`
    b += 1;

    dbg!(a);
    dbg!(b);
}
```

We can call the function `mutability()` with the following code:

```
let a = 1;
let b = 2;                    The variable b is immutable, but it's moved
mutability(a, b);            into the function as a mutable variable.
```

Running this code produces the following output:

```
[src/main.rs:8] a = 1
[src/main.rs:9] b = 3
```

Note that in this example, we're switching the mutability of b from immutable to mutable when we pass it to the function `mutability()` simply by applying the `mut` keyword to the argument. In some cases, doing so can confuse the function caller, but because the ownership is transferred (moved) to the function, the function can do whatever it wants with the value. Although we're altering the mutability of b, the change doesn't affect the original variable b in the caller's scope, so the pattern isn't likely to be dangerous.

It's impossible to change the mutability of a reference in the same way we can with owned values across a function call. References have been borrowed, and you cannot change the mutability of borrowed data (the entire raison d'être of Rust's borrow checker).

NOTE Remember one more important thing about passing arguments by value in Rust: the compiler will invoke the `Copy` trait when it's available. `Copy` differs from `Clone` in that `Copy` is a marker trait that tells the compiler that the type can be copied by copying the bits in memory, whereas `Clone` is a trait that provides a method to clone a value explicitly. The implication is that any value that implements `Copy` will be copied rather than moved when passed to a function or assigned to a new variable. In most cases, this applies only to primitive types such as integers, floats, and Booleans, but it can also apply to types such as tuples, arrays, and structs that contain only `Copy` types.

We can also use `RefCell` to achieve interior mutability.

Listing 9.3 Using `RefCell` for interior mutability

```
let immutable_string =
    String::from("This string cannot be changed");      Declares an immutable string
// immutable_string.push_str("... or can it?"); // error: cannot borrow
`immutable_string` as mutable, as it is not declared as mutable
dbg!(&immutable_string);

let not_so_immutable_string = RefCell::from(immutable_string);
not_so_immutable_string
    .borrow_mut()                          Creates a RefCell from the immutable
    .push_str("... or can it?");           string, which moves the string into the
dbg!(&not_so_immutable_string);            RefCell. Note that the RefCell is not
                                           declared as mutable.
```

Now we can mutate the
string inside the RefCell.

Uncommenting this line will
result in a compilation error.

Running this code produces the following output:

```
[src/main.rs:32] &immutable_string = "This string cannot be changed"
[src/main.rs:38] &not_so_immutable_string = RefCell {
    value: "This string cannot be changed... or can it?",
}
```

As you can see, Rust's shared mutable containers provide a bypass for the mutability rules. It's important to remember that the `RefCell` owns its data. In the preceding example, we moved the string into the `RefCell`, which involves a plain old function call that allows us to alter the mutability of any owned value.

9.6   *Using traits to make (almost) anything immutable*

We've discussed some of the benefits and downsides of immutability, but we need to explore how to put it into practice. Rust's standard library provides a few tools to help us. In this section, we'll discuss the `std::borrow::ToOwned` trait, which gives us the basis for a pattern that we can use to make just about anything immutable.

   When we're working with immutable data, we want to avoid making copies of data when we don't need to. To do so, we use references to borrowed data. You may have seen code in Rust that looks like this example:

```
let s = "A static string".to_owned();
```

This code uses the `ToOwned` trait to convert a `&str` to a `String`. Rust provides a blanket implementation for any type `T`, where `T` provides `Clone`. In other words, we can think about `ToOwned` as being a generalization of `Clone` for references or slices. The implementation of `ToOwned::to_owned()` simply calls `Clone::clone()`, and for `[T]`, it returns a `Vec` with each item cloned. The following listing shows the definition of the `ToOwned` trait.

Listing 9.4   Definition of `ToOwned` from Rust's standard library

```
pub trait ToOwned {
    type Owned: Borrow<Self>;

    // Required method
    fn to_owned(&self) -> Self::Owned;

    // Provided method
    fn clone_into(&self, target: &mut Self::Owned) { ... }
}
```

Knowing this definition, we don't have to do much to use immutable data everywhere. All we have to do is provide a `Clone` implementation; then we can use `ToOwned` to convert our data to an owned value when we need to mutate it. This process may seem a bit clunky, but we can use the `Cow` type to make it more ergonomic.

9.7 *Using Cow for immutability*

The Cow type is a smart pointer that implements a clone-on-write pattern. Cow itself is implemented as an enum, which requires the ToOwned trait to be implemented for its contents. You've likely heard of copy-on-write, and Cow follows the same pattern by deferring the cost of cloning until it's necessary. We can use Cow as a container for data that we want to treat as immutable by mutating only cloned data, never the source data. If we're being nitpicky, this approach doesn't strictly prevent mutability, but it prevents us from having to mutate the source data or use any mutable references. The following listing shows the definition of Cow.

> **Listing 9.5 Definition of Cow from Rust's standard library**

```
pub enum Cow<'a, B>
where
    B: 'a + ToOwned + ?Sized,
{
    Borrowed(&'a B),
    Owned(<B as ToOwned>::Owned),
}
```

Notice a couple of things about Cow:

- The Cow type is generic over a lifetime 'a and a type B, which must implement the ToOwned trait.
- The Cow type is an enum with two variants: Borrowed and Owned. It's similar to an Option but more specialized.
- We can use Cow to wrap any reference for a type that implements Clone (and thereby ToOwned) and then obtain an owned value when we need to mutate it.
- Cow implements the Deref trait, which allows us to treat it as a reference to the data it contains.

The following listing shows the basic use of Cow.

> **Listing 9.6 Basic use of Cow**

```
use std::borrow::Cow;

let cow_say_what = Cow::from("The cow goes moo");
dbg!(&cow_say_what);

let cows_dont_say_what =
    cow_say_what
        .clone()
        .to_mut()
        .replace("moo", "toot");
dbg!(&cow_say_what);
dbg!(&cows_dont_say_what);
```

We can mutate the cloned data without affecting the source data. Note, however, that we need to clone the Cow and then call to_mut() to obtain a mutable reference.

The source data is still immutable.

The cloned data was mutated.

Notice that we still need to call `clone()` to obtain a new `Cow`, which is for the smart pointer itself, not the data it contains. Then we call `to_mut()` to obtain a mutable reference to the internal data after it's cloned. Running the preceding code produces the following output:

```
[src/main.rs:5:5] &cow_say_what = "The cow goes moo"
[src/main.rs:9:5] &cow_say_what = "The cow goes moo"
[src/main.rs:10:5] &cows_dont_say_what = "The cow goes toot"
```

Let's try to improve on that example to clarify how you'd use it in practice. Let's write a function that does something similar: returns a new object.

Listing 9.7   Improving the use of `Cow`

```
fn loud_moo<'a>(mut cow: Cow<'a, str>)
        -> Cow<'a, str> {
    if cow.contains("moo") {
        Cow::from(cow.to_mut().replace("moo", "MOO"))
    } else {
        cow
    }
}
```

The function takes an owned Cow and returns a Cow.

If the Cow contains "moo", we mutate it and replace it with "MOO".

If the Cow doesn't contain "moo", we return the original Cow.

We can call the function `loud_moo()` with the following code:

```
let cow_say_what = Cow::from("The cow goes moo");
let yelling_cows = loud_moo(cow_say_what.clone());
dbg!(&cow_say_what);
dbg!(&yelling_cows);
```

When we run the code, we get the following output:

```
[src/main.rs:21:5] &cow_say_what = "The cow goes moo"
[src/main.rs:22:5] &yelling_cows = "The cow goes MOO"
```

If we're using `Cow`, we don't necessarily want to leak that implementation detail in a public API, so we'd likely want to wrap our data with a struct. We can put a `Cow` inside a struct and provide a method to mutate the internal data without exposing the `Cow` itself.

Listing 9.8   Wrapping `Cow` in a struct

```
#[derive(Debug, Clone)]
struct CowList<'a> {
    cows: Cow<'a, [String]>,
}

impl<'a> CowList<'a> {
    fn add_cow(&self, cow: &str) -> Self {
        let mut new_cows = self.clone();
```

We derive Clone so that we can clone the CowList, including the internal cow list.

We use Cow to wrap a vector of strings, including the 'a lifetime.

We provide a method to add a cow to the list, returning a new CowList.

We clone the CowList first so that we can mutate it.

```
        new_cows.cows.to_mut().push(
            cow.to_string()
        );
        new_cows
    }
}

impl Default for CowList<'_> {
    fn default() -> Self {
        CowList {
            cows: Cow::from(vec![]),
        }
    }
}
```

> We mutate the internal Cow by calling to_mut() and then push().

> We return the new CowList.

Now let's test our code:

```
let list_of_cows = CowList::default()
    .add_cow("Bessie")
    .add_cow("Daisy")
    .add_cow("Moo");
dbg!(&list_of_cows);
let list_of_cows_plus_one = list_of_cows.add_cow("Penelope");
dbg!(&list_of_cows);
dbg!(&list_of_cows_plus_one);
```

> The original CowList is still immutable, as we can see by printing it twice.

Running this code produces the following output:

```
[src/main.rs:49:5] &list_of_cows = CowList {
    cows: [
        "Bessie",
        "Daisy",
        "Moo",
    ],
}
[src/main.rs:52:5] &list_of_cows = CowList {
    cows: [
        "Bessie",
        "Daisy",
        "Moo",
    ],
}
[src/main.rs:53:5] &list_of_cows_plus_one = CowList {
    cows: [
        "Bessie",
        "Daisy",
        "Moo",
        "Penelope",
    ],
}
```

As an alternative implementation, we can place each `Cow` inside a `Vec`:

```
#[derive(Debug, Clone)]
struct CowVec<'a> {
    cows: Vec<Cow<'a, str>>,
}
```

One advantage of using this method (as opposed to using a `Vec` within a `Cow`) is that each item in the vector can be cloned lazily, which is known as *structural sharing*. This approach can be more efficient when you have many copies of the same elements, especially if the individual elements are large.

As you can see, `Cow` isn't complicated, but we can hide it from a public API to make our API a little easier to use. Note that in the preceding example, we never alter the original or source `CowList` and always return a new `CowList` when we mutate it.

We can apply the use of `Cow` nearly anywhere we want to encourage the use of immutable data, but we still need to understand `Cow` and its behavior. If you haven't encountered `Cow` yet, it may seem to be an odd abstraction, especially when you can call `clone()` directly or mutate the data. If you think that this approach is an awkward way to work with data, you're not alone, which is why the next section discusses some data structures that make applying immutability a bit easier.

9.8 Using crates for immutable data structures

In this section, we'll explore some crates that provide immutable data structures, which can be a relatively easy way to reap the benefits of immutability without building custom solutions. The crates we'll discuss are

- `im`, which provides lists, sets, and maps (https://crates.io/crates/im)
- Rust Persistent Data Structures (`rpds`), which provides lists, sets, queues, and maps (https://crates.io/crates/rpds)

Both crates provide structures optimized for use in programs and libraries that follow the principles of immutability, but neither strictly enforces immutability through their APIs. You can use the data structures in these crates in a mutable way, but they are optimized for immutability.

9.8.1 Using im

The `im` crate is the most popular library that provides immutable data structures. `im` has well over 7.4 million downloads from https://crates.io and is used in many other crates and projects.

`im` provides a `Vector` analogous to Rust's `Vec`, which is optimized for immutability. It also has o ordered and unordered sets and hash maps, each tuned for immutability. We can use `im` to create a `Vector` with the `vector!` macro and add elements to it.

> **Listing 9.9 Using `im` to create a `Vector`**

```
use im::vector;

let shopping_list =
    vector!["milk", "bread", "butter", "cheese", "eggs"];
```

```
let mut updated_shopping_list = shopping_list.clone();
updated_shopping_list.push_back("grapes");

dbg!(&shopping_list);
dbg!(&updated_shopping_list);
```

We need to clone the original vector.

Note that we mutate the vector like a normal vector and append an element mutably with push_back().

Something to note about the `Vector` from `im` is that using it mutably is possible; it doesn't *force* us to make a copy of the vector each time. `im` is intended to be used as the underlying immutable data structure without enforcing immutability. Remember that the `Clone` implementation provided is optimized for the immutability use case, so we can clone the `Vector` liberally without worrying about performance. Running this code produces the following output:

```
[src/main.rs:10:5] &shopping_list = [
    "milk",
    "bread",
    "butter",
    "cheese",
    "eggs",
]
[src/main.rs:11:5] &updated_shopping_list = [
    "milk",
    "bread",
    "butter",
    "cheese",
    "eggs",
    "grapes",
]
```

In addition to data structures such as sets and maps, `im` provides features such as rayon-based iterators (for parallel iterating), Serde support, and proptest and quick-check support for property-based testing. For complete details, consult the `im` documentation at https://docs.rs/im/latest/im.

9.8.2 *Using rpds*

The `rpds` crate is similar to `im` but provides a few additional data structures, such as queues and stacks. Although `rpds` is less popular than `im`, at around 1.2 million downloads from https://crates.io, it's still a well-maintained and useful library.

Unlike `im`, `rpds` provides an immutable API directly (methods that return a new structure), although it also provides a mutable API if you want to avoid cloning. You can use `rpds` to create a `Vector` and add elements to it.

Listing 9.10   Using `rpds` to create a `Vector`

```
use rpds::Vector;

let streets = Vector::new()
    .push_back("Elm Street")
```

```
        .push_back("Maple Street")
        .push_back("Oak Street");

let updated_streets = streets.push_back("Pine Street");

dbg!(&streets);
dbg!(&updated_streets);
```

Note that with rpds, each call to push_back() returns a new Vector, so you don't have to clone explicitly. It also provides a push_back_mut() if you need to modify the vector in place. If you run the code, you get the following output:

```
[src/main.rs:11:5] &streets = Vector {
    root: Leaf(
        [
            "Elm Street",
            "Maple Street",
            "Oak Street",
        ],
    ),
    bits: 5,
    length: 3,
}
[src/main.rs:12:5] &updated_streets = Vector {
    root: Leaf(
        [
            "Elm Street",
            "Maple Street",
            "Oak Street",
            "Pine Street",
        ],
    ),
    bits: 5,
    length: 4,
}
```

rpds also provides Serde support and macros for initialization data structures. For complete details, consult the rpds documentation at https://docs.rs/rpds/latest/rpds.

Summary

- Immutability is a powerful abstraction for writing reliable software.
- Immutability can help prevent logic errors, race conditions, unwanted side effects, and memory safety problems.
- We can combine immutability with functional programming patterns, such as pure functions and referential transparency, to make our code more reliable, easy to test, and easier to reason about.
- Rust always distinguishes between mutable and immutable values, enforced by the borrow checker, which makes it easy to determine whether a value is mutable thanks to inherited mutability.

- Rust provides a few tools to help us work with immutable data, such as the `ToOwned` trait and the `Cow` type.
- The `im` and `rpds` crates provide data structures optimized for immutability, which can be used as a building block for programs and libraries that use immutable data.

Antipatterns

This chapter covers

- Discussing programming antipatterns
- Reviewing common antipatterns in Rust
- Recognizing when to use and when to avoid contentious patterns

Antipatterns are programming practices that are considered harmful in specific contexts or all circumstances. Antipatterns are often the result of a misunderstanding of languages or a lack of experience with a particular technology stack. In this chapter, we'll discuss some common antipatterns in Rust and how to avoid them.

First, we must discuss what constitutes an antipattern and then explore Rust's most common examples. We'll also discuss when to use—and when to avoid—specific patterns and when to make exceptions.

The rules presented in this chapter aren't hard-and-fast rules; exceptions always exist. But it's important to understand the reasoning behind these rules and know when to break them. As Rust evolves, these rules may change, so staying up to date with the latest best practices is essential for writing Rust effectively.

205

10.1   *What is an antipattern?*

Antipattern is a bit of a weasel word. That is, it's often used pejoratively to refer to any practice that the speaker doesn't like. Ultimately, the definition of *antipattern* is a matter of opinion and preference. In some cases, however, a practice is objectively bad, such as when it's unsafe, inefficient, or difficult to maintain. These cases likely arise from a combination of bad design, a desire to maintain backward compatibility, and a continuously changing landscape of acceptable software design.

The C language offers a great case study of how language-design practices evolve. The C language is arguably the single most influential programming language in history. Despite its ubiquity, C is also arguably one of the worst languages in terms of safety and ease of use, particularly in systems programming. Even highly skilled experts can easily make mistakes in C that are difficult to detect and correct.

Some people might argue that the C language is objectively bad by modern standards, and I have to agree. I've spent many hours dealing with bugs in C that are easy to make but hard to catch due to the language's design. Writing C can be nostalgic in that it's like driving a Ford Model T—fun for a while but not something you'd want to do every day. Nevertheless, the C language remains the best choice for many applications today, particularly in systems and embedded programming or when the next-best alternative is assembly.

The Rust language was designed carefully and thoughtfully to avoid the footguns you find in a language like C. Rust also attempts to preempt one compelling argument in favor of C by being just as fast, if not faster, than C. In many benchmarks, Rust outperforms C in terms of raw speed and does so without unsafe code.

Even Rust, however, has become a victim of its own success in that its popularity has made it difficult to make significant changes. Changes that would break backward compatibility are challenging to argue for and make because the cost of rewriting existing code is high enough that people will avoid upgrading (a problem that C and other languages have had for decades).

Comparing Rust with C is a bit unfair because Rust is possible only thanks to significant improvements in compiler infrastructure (namely, the LLVM project; https:// llvm.org) and a better understanding of programming-language design than we had in the 1970s. What we may consider an antipattern in C today may have been a best practice in the 1970s. The cool thing about Rust is that the compiler does much of the work for you, provided that you avoid using `unsafe` code blocks. The same cannot necessarily be said of languages like C, in which the compiler is somewhat of a blunt instrument and leaves much of the optimization to the programmer. (Strictly speaking, though, both Clang and GCC do an excellent job of optimizing C code.)

Nevertheless, the topic of antipatterns in Rust is worth exploring, and I hope that you walk away from this chapter with a better understanding of Rust and its limitations. Finally, to answer the question "What's an antipattern?", I simply define an antipattern as any pattern you don't like.

10.2 Using unsafe

The mother of all antipatterns in Rust is the inappropriate use of the dangerous `unsafe` keyword. You have to use the `unsafe` keyword to accomplish many things in Rust, but it's also the best way to shoot yourself in the foot. You can think of `unsafe` as an escape hatch that allows you to perform operations that violate Rust's language rules, such as working with raw pointers, calling C functions, and accessing or modifying resources outside a program's allocated memory space. In the vast majority of use cases in Rust, you shouldn't need the `unsafe` keyword, and you should scrutinize any use of it.

That said, it's nearly impossible to use Rust without using `unsafe` code, at least indirectly, because the standard library uses `unsafe` code throughout. You needn't look hard through the standard library's code to find uses of `unsafe`, such as in the implementations of `Box`, `Vec`, and `String`. Memory allocation and deallocation, OS system calls, and other low-level operations are also `unsafe` operations. Many examples of `unsafe` code in the standard library are either optimizations or necessary operations that can't be performed safely otherwise (C-style foreign function interface [FFI], system calls, and so on). The implementation of the `Vec::insert()` method, for example, includes the block of `unsafe` code shown in the following listing, which provides an optimized implementation of insertion within a vector.

Listing 10.1 `Vec::insert()` **from the Rust standard library**

```
pub fn insert(&mut self, index: usize, element: T) {
    #[cold]
    #[cfg_attr(not(feature = "panic_immediate_abort"), inline(never))]
    #[track_caller]
    fn assert_failed(index: usize, len: usize) -> ! {
        panic!("insertion index (is {index}) should be <= len (is {len})");
    }

    let len = self.len();

    // space for the new element
    if len == self.buf.capacity() {
        self.reserve(1);
    }
                                          ┌── The start of the
    unsafe {                          ◄───┘   unsafe block
        // infallible
        // The spot to put the new value
        {
            let p = self.as_mut_ptr().add(index);   ◄──
            if index < len {
                // Shift everything over to make space. (Duplicating the
                // `index`th element into two consecutive places.)
                ptr::copy(p, p.add(1), len - index);
            } else if index == len {
                // No elements need shifting.
            } else {
```

Gets a mutable pointer to the Vec's buffer, using as_mut_ptr() and pointer arithmetic. The call to as_mut_ptr() is safe, but the call to add() on the returned pointer is unsafe.

Shifts elements to make space for the new element, using pointer arithmetic with ptr::copy()

```
            assert_failed(index, len);
        }
        // Write it in, overwriting the first copy of the `index`th
        // element.
        ptr::write(p, element);        ◁────┐  Writes the new
    }                                        │  element into the Vec
    self.set_len(len + 1);                   │  using ptr::write()
}
}
```

You may wonder why the `insert()` method is implemented with an `unsafe` code block. The short answer is that the unsafe version of insertion is much faster than the safe version, so the authors of the standard library chose to make this tradeoff for performance reasons. Notably, `ptr::copy()` is equivalent to C's `memmove()`, which can be optimized with SIMD (single instruction, multiple data) and other low-level instructions as long as the memory regions being copied don't overlap. The code is written to be infallible, meaning that the code can't panic or cause undefined behavior even though it's technically unsafe. In listing 10.1, the author of the standard library did a good job of documenting the use of `unsafe` by providing comments about its use (which is good practice).

10.2.1 What does unsafe do?

Let's take a moment to understand why `unsafe` is necessary in Rust and when you might need to use it. The `unsafe` keyword in Rust has the following effects:

- It allows you to dereference raw pointers.
- It permits you to call unsafe functions or methods.
- It allows you to access or modify mutable static variables.
- It allows you to implement unsafe traits.
- It allows you to access fields of `union` types, which are provided for C compatibility.

In practice, you'll need to use `unsafe` most frequently when you're working with C libraries or other FFI-based code. If you want to integrate with a Python library, for example, you must use `unsafe` to call the Python C API. But you'd likely be better off using a framework such as PyO3 (https://github.com/PyO3/pyo3) that provides the necessary bindings.

C libraries in particular are known for their use of raw pointers, manual memory allocation and deallocation, and other no-nos in Rust. You need to use C libraries when you're working with system calls, which you must do to interact with the OS (reading and writing files, creating and managing processes, accessing peripherals, and so on).

The Rust standard library provides many safe abstractions, so you don't need to write unsafe code yourself, but you should be aware that you're using unsafe code when you use these abstractions. The `std::ffi` module, for example, provides safe abstractions over FFI, such as `CString`, `CStr`, `OsStr`, and `OsString`. The `std::fs` module provides safe abstractions over file I/O, such as `File`, `DirEntry`, and `Metadata`. The `std::process` module provides safe abstractions over process management, such as `Command`, `ExitStatus`, and `Stdio`.

For pointer arithmetic, the `std::ptr` module provides abstractions that allow you to work with pointers. Most of the key methods are unsafe, however, thus requiring the use of `unsafe` blocks.

Memory allocation and deallocation are also unsafe operations, and the standard library provides safe abstractions over these operations, such as `Box`, `Vec`, and `String`. Under the hood, `Box`, `Vec`, and `String` use Rust's `Allocator` trait, which wraps `malloc()` and `free()` in UNIX-like systems and `HeapAlloc()` and `HeapFree()` in Windows. The allocator API is a set of unsafe functions that are part of the unsafe `Allocator` trait, which allows you to allocate and deallocate memory, and the standard library uses it to provide safe abstractions over memory allocation and deallocation. The allocator API is still experimental and available only in nightly Rust.

10.2.2 Where can you use unsafe?

You can use `unsafe` in the following ways:

- You can define a code block as `unsafe` by using the `unsafe` keyword, wrapped by braces, such as `unsafe { … }`. The block is evaluated as an expression, and the value of the block is the value of the last expression in the block.
- You can define a function as `unsafe` by using the `unsafe` keyword, such as `unsafe fn foo() { … }`. You can call an unsafe function only from within an `unsafe` block or another unsafe function.
- You can define a trait as `unsafe` by using the `unsafe` keyword, such as `unsafe trait Foo { … }`. An unsafe trait can contain safe and unsafe methods, but you can call an unsafe method only from within an `unsafe` block or function. Any trait with one or more unsafe methods is considered unsafe.

The following listing demonstrates an `unsafe` function by calling `printf()` from the C standard library.

Listing 10.2 Defining an `unsafe` function to call `printf()`

```
unsafe fn unsafe_function() {
    libc::printf(
        "calling C's printf() within unsafe_function()\n\0".as_ptr()
            as *const i8,
    );
}
```

You can test this code by calling the `unsafe_function()` from within an `unsafe` block.

Listing 10.3 Calling `unsafe_function()` from within an `unsafe` block

```
unsafe {
    unsafe_function();
}
```

Running this code will produce the following output:

```
calling C's printf() within unsafe_function()
```

You can also define an unsafe trait.

Listing 10.4   Defining an unsafe trait

```
unsafe trait UnsafeTrait {
    fn safe_method(&self);
    unsafe fn unsafe_method(&self);
}

struct MyStruct;

unsafe impl UnsafeTrait for MyStruct {
    fn safe_method(&self) {
        println!("calling println!() within UnsafeTrait::safe_method()");
    }
    unsafe fn unsafe_method(&self) {
        libc::printf(
            "calling C's printf() within UnsafeTrait::unsafe_method()\n\0"
                .as_ptr() as *const i8,
        );
    }
}
```

This example also has a safe method, safe_method(), which you can call without using an unsafe block. You can call the unsafe_method() from within an unsafe block.

Listing 10.5   Testing an unsafe trait

```
let my_struct = MyStruct;
my_struct.safe_method();
unsafe {
    my_struct.unsafe_method();
}
```

Running this code will produce the following output:

```
calling println!() within UnsafeTrait::safe_method()
calling C's printf() within UnsafeTrait::unsafe_method()
```

You may notice that it's possible to hide unsafe code behind safe abstractions; whether that capability is a feature or a bug is a matter of opinion. In practice, it's impossible to create a programming language like Rust that's 100% free of unsafe code, so Rust's choice to allow safe code to hide unsafe code is a pragmatic one.

You can use the #![forbid(unsafe_code)] attribute to ensure that your crate doesn't contain unsafe code, but the attribute doesn't apply to crates you include as dependencies or to the Rust standard library. In other words, even when you use the

`#![forbid(unsafe_code)]` attribute, you're very likely to be using unsafe code, even if you don't write it yourself.

Listing 10.6 Using `#![forbid(unsafe_code`

```
#![forbid(unsafe_code)]

fn main() {
    // unsafe {
    //     libc::printf("Hello, world!\n".as_ptr() as *const _);
    // }
    let mut fruits = vec!["apple", "banana", "cherry"];
    fruits.insert(0, "orange");
}
```

Uncommenting this line will cause a compilation error.

This line will not cause a compilation error even though it calls Vec::insert(), which contains unsafe code.

NOTE At the time of this writing, Rust doesn't provide a way to ensure that dependencies are free of unsafe code, but the `cargo-geiger` crate (https://crates.io/crates/cargo-geiger) can be used to analyze the amount of unsafe code in a crate and its dependencies.

10.2.3 When should you use unsafe?

The main use cases for unsafe code are as follows:

- Working with C libraries or other FFI-based code
- Making system calls that don't have safe abstractions in the standard library
- Implementing safe abstractions over unsafe code
- Writing low-level optimizations that can't be expressed safely

Some Rustaceans are dogmatic about avoiding unsafe code, but a more pragmatic view (which I share) is that we should avoid `unsafe` when possible but not be afraid to use it when necessary. When you do need to use `unsafe`, you need to take extra care to ensure that your code is correct and doesn't cause undefined behavior. This is easier said than done, and it's one of the reasons why unsafe code is considered an antipattern.

When using unsafe code is unavoidable, you can reduce the likelihood of introducing critical bugs by using robust tools such as property testing, fuzz testing, and static analysis tools. The Rust community has developed a set of guidelines for `unsafe` use, which you can find at https://rust-lang.github.io/unsafe-code-guidelines.

10.2.4 Should you worry about unsafe?

For the most part, you don't need to worry about unsafe code, particularly in the Rust standard library. The standard library is well tested and well maintained, and the Rust core team is vigilant about ensuring that the standard library is free of undefined behavior. The standard library is also designed to provide safe abstractions over unsafe code, so you don't need to write unsafe code in most cases.

I've encountered cases in which the use of unsafe code was necessary, such as working with OS-level abstractions that are not fully covered by the standard library. One downside to the standard library's abstractions is that they're designed to be cross-platform, and they generally represent the lowest common denominator of what's possible on all platforms. As a result, you may need to use unsafe code to access platform-specific features such as the Windows API or to take advantage of platform-specific optimizations. I've found that unsafe code isn't as scary as it's made out to be if you have a good grasp of Rust, the borrow checker, and best practices for managing resources in Rust, such as using resource acquisition is initialization (RAII) and smart pointers.

10.3   *Using unwrap()*

Improper use of the `unwrap()` method is a common antipattern in Rust that's often used when we get lazy about handling `Option` or `Result` values. But you can avoid using `unwrap()` relatively easily by replacing it with one or more of the following methods:

- `expect()`—This method is similar to `unwrap()` but allows you to provide a custom error message when the value is `None` or `Err` (for `Option` and `Result`, respectively). `expect()` can be useful for debugging, but using it to handle errors in production code is a good idea only when the expected behavior is that the program should exit. Using `expect()` is functionally equivalent to using an assertion, such as `assert!(value.is_some())`.
- `map()`—This method allows you to transform the value of an `Option` or `Result` by using a closure. If the value is `None` or `Err`, the closure is not called, and the method returns `None` or `Err`.
- `and_then()`—This method allows you to chain `Option` or `Result` values, avoiding deeply nested `match` or `if let` statements.
- `unwrap_or()`—This method allows you to provide a default value when the value is `None` or `Err` and prevents panic.
- `?`—This operator allows you to propagate errors up the call stack and is particularly useful when you're working with `Result` values.

`unwrap()` isn't always an antipattern, but it's often a code smell, as it can indicate that you're not thinking about error handling or the possibility of `None` values. It's also a sign that you're not thinking about the control flow of your program or the possibility of failure.

There are exceptions, such as when you're reasonably certain that a value will never be `None` or `Err`. In these cases, it's much better to use `expect()` with a custom error message, as it will provide more information when the value is `None` or `Err`.

10.4   *Not using Vec*

The `Vec` type, a dynamic array that's one of the most commonly used types in Rust, is a good choice for most use cases. Many people make the mistake of not using `Vec`, instead attempting to optimize their code by writing custom data structures or reaching for maps, sets, trees, or linked lists.

As it turns out, `Vec` is remarkably fast for many workloads. In many cases, it provides the best performance when you account for a variety of benchmarks. A `HashSet` or `HashMap`, for example, has exceptionally quick lookups, but if you need to append new elements to the collection, `Vec` is often faster. The same is true of `BTreeSet` and `BTreeMap`, which are great for ordered collections but not as fast as `Vec` for many workloads. `LinkedList` is often slower than `Vec` for many workloads and also less memory-efficient. To demonstrate, I've written a simple set of benchmarks for `Vec`, `HashSet`, and `LinkedList` that perform the following operations:

- Appending 1,000,000 elements to an empty collection
- Finding 1,000 random values within a collection of 1,000,000 unique elements
- Removing 1,000 elements from a collection of 1,000,000

The following listing shows the `append` benchmark. (Refer to the book's source code for the full benchmark.)

Listing 10.7 Benchmarking `Vec`, `HashSet`, and `LinkedList` for appending

```
#[bench]
fn vec_append(b: &mut Bencher) {
    b.iter(|| {
        let mut nums: Vec<i32> = Vec::new();
        for n in 0..1_000_000 {
            nums.push(n);
        }
    });
}

#[bench]
fn list_append(b: &mut Bencher) {
    b.iter(|| {
        let mut nums: LinkedList<i32> = LinkedList::new();
        for n in 0..1_000_000 {
            nums.push_back(n);
        }
    });
}

#[bench]
fn set_append(b: &mut Bencher) {
    b.iter(|| {
        let mut nums: HashSet<i32> = HashSet::new();
        for n in 0..1_000_000 {
            nums.insert(n);
        }
    });
}
```

When we run all the benchmarks, we find that although `Vec` is not always fastest, it performs surprisingly well for all three tests. The following listing shows the results.

> **Listing 10.8   Benchmark results for `Vec`, `HashSet`, and `LinkedList`**

```
running 9 tests
test tests::list_append ... bench:  53,860,800 ns/iter (+/- 2,306,429)
test tests::list_find    ... bench:     527,207 ns/iter (+/- 26,305)
test tests::list_remove ... bench:  61,830,454 ns/iter (+/- 1,462,953)
test tests::set_append  ... bench:  23,774,245 ns/iter (+/- 549,095)
test tests::set_find     ... bench:          11 ns/iter (+/- 0)
test tests::set_remove  ... bench:     839,977 ns/iter (+/- 4,571)
test tests::vec_append  ... bench:   2,095,262 ns/iter (+/- 146,611)
test tests::vec_find     ... bench:     133,359 ns/iter (+/- 11,424)
test tests::vec_remove  ... bench:   3,319,558 ns/iter (+/- 57,979)

test result: ok. 0 passed; 0 failed; 0 ignored; 9 measured; 0 filtered out;
finished in 136.97s
```

NOTE Running these benchmarks on your machine may produce different results. The benchmarks use a nightly-only benchmarking feature, and to run these benchmarks, you must use the `cargo bench` command (as opposed to `cargo test`).

`Vec` beats `LinkedList` on every benchmark, and `HashSet` is faster for removing and finding elements but significantly slower for appending new elements. The `Vec` type is more memory-efficient than `HashSet` and `LinkedList`, and it's easier to work with in many cases.

In terms of complexity, these results aren't far from what we expect. Table 10.1 shows the big O and big theta complexity for common operations with `Vec`, `HashSet`, and `LinkedList`.

NOTE My analysis differs from what you'll find in the Rust documentation because the documentation doesn't discern between average and worst-case complexity.

Table 10.1   Summary of big O and big theta complexity for common operations with `Vec`, `HashSet`, and `LinkedList`

Structure	Append		Search		Remove	
	Average	Worst	Average	Worst	Average	Worst
`Vec`	$\Theta(1)$	$O(n)$	$\Theta(n)$	$O(n)$	$\Theta(n)$	$O(n)$
`HashSet`	$\Theta(1)$	$O(n)$	$\Theta(1)$	$O(n)$	$\Theta(1)$	$O(n)$
`LinkedList`	$\Theta(1)$	$O(1)$	$\Theta(n)$	$O(n)$	$\Theta(n)$	$O(n)$

`Vec` doesn't appear to perform remarkably well in any operation other than indexed lookups, which are $O(1)$ (not in the table or benchmarks). But the average performance in practice is surprisingly good under various workloads. Confusingly, `LinkedList` is significantly worse than `Vec` for inserting 1 million elements one at a time, but this poor performance likely occurs because `LinkedList` has to allocate memory for each element, whereas `Vec` allocates memory in chunks. `HashSet`'s performance in append or insert operations is also poor due to allocations and the cost of rebalancing the hash table as it grows.

Rust's benchmarking tools

Rust provides a built-in benchmarking tool that allows you to write benchmarks quickly the way you'd write a unit test. Currently, this feature is available only in nightly Rust.

Using the `#[bench]` attribute, you can define a unit test that benchmarks any operation, like any regular unit test. The benchmarking tool runs the benchmark multiple times and provides the median time to run the benchmark and the standard deviation.

Rust's test library includes the `Bencher` object, which provides a method for measuring the time it takes to run a block of code. The `Bencher` struct provides an `iter()` method that accepts a closure, in which you can place the code you want to benchmark. Any setup or teardown should occur before and after the call to `Bencher::iter()`. A minimal benchmarking test looks like the following code:

```
#![feature(test)]

#[cfg(test)]
mod test {
    extern crate test;
    use test::Bencher;
    #[bench]                                    ← The #[bench] attribute
    fn hello_world_10_times(b: &mut Bencher) {     marks the function as a
        b.iter(|| {                       ←        benchmark.
            for _ in 0..10 {              ←     The iter() method of the
                println!("Hello, world!");          Bencher object runs the
            }                                       benchmark multiple times.
        });
    }            The code to be benchmarked is
}                placed inside the closure passed to
                 iter(). Note that we are running the
                 test 10 times in this example within
                 the closure, which will also be
                 called multiple times.
```

Running the `cargo bench` command compiles the code in release mode and executes the benchmarks. `cargo bench` takes arguments similar to the `cargo test` command, allowing you to filter benchmarks by name or run only specific benchmarks. When you run `cargo bench`, Rust's test library runs the code within `Bencher::iter()` per the following rules to obtain a stable result:

1 The benchmark is run 50 times, and a summary of the results is calculated.
2 The outliers are removed from the results (the fastest and slowest 5% of the results).
3 The benchmark is run again 50 times, and the results are calculated.

(continued)

   4 If either of the following conditions is met, the results are returned:
      – The standard deviation of the results is less than 100 milliseconds.
      – The benchmark has been running for more than 3 seconds.
   5 If neither condition is met, the benchmark runs again from step 1.

If you want to run benchmarks in stable Rust, you can use the `Criterion.rs` crate (documented at https://bheisler.github.io/criterion.rs/book), which provides a feature-rich benchmarking tool. `Criterion.rs` is a Rust port of Haskell's Criterion library.

`Vec` benefits from being a contiguous block of memory, which makes it cache-friendly in most modern CPUs and gives the compiler opportunities to optimize operations at the instruction level. Data locality is a key performance factor, especially when accessing the main memory (RAM) on computers is orders of magnitude slower than accessing the CPU's caches. `Vec` also benefits from the relative simplicity of managing a contiguous block of memory. Shifting elements around is relatively straightforward and doesn't require complex algorithms; in most cases, it's merely a matter of copying memory, which can be extremely fast on modern computers.

Indeed, in some cases, a set, map, tree, or linked list will handily outperform a vector, but you may find it harder to justify using these types than you think. `Vec` is a good choice for most workloads and often an excellent choice for many workloads. When in doubt, use `Vec`, or at least take time to benchmark your code before reaching for a more complex data structure.

10.5 *Too many clones*

Some Rustaceans cringe at the sight of the `clone()` method, and in many cases, they have good reason to do so. Although I'm not an anticlone zealot, I do think that the `clone()` method is often overused and used when it's not necessary.

The `clone()` method creates a deep copy of a value, and from what I've seen, some Rust programmers use it as a crutch to avoid thinking about ownership and borrowing. This approach is a mistake: it can lead to performance problems and memory bloat, and it can also cause bugs.

Calling `clone()` isn't always bad, however. In chapter 9, I advocate for the use of `clone()` as a simple way to implement immutable data structures. If you find yourself using `clone()` to bypass or get around the borrow checker, you should take a step back and think about your design. As with everything, though, if you're making choices that are informed and deliberate, you shouldn't feel bad about using `clone()`—especially if your decisions are based on benchmarks and data.

10.6   *Using Deref to emulate polymorphism*

Polymorphism is a technique that allows you to treat objects of different types as though they're the same type. Object-oriented languages encourage the use of polymorphism through subtyping or inheritance, which are notably absent from Rust.

Sometimes, we use the `Deref` trait to make it easier to work with objects by letting the compiler infer the methods we want to call by using `Deref` coercion. In a way, we're effectively emulating the kind of polymorphism you may have seen in other languages, such as C++ and Java. This approach isn't necessarily bad, but it can be a sign that we're not thinking about our design in a way that's idiomatic to Rust.

The `Deref` trait (and its mutable counterpart `DerefMut`) allows us to dereference a value by using the `*` operator, as in `*value`. Also, the compiler implicitly uses the `Deref` trait to allow method calls on a value that's wrapped in a smart pointer, such as `Box`, `Rc`, or `Arc`. In other words, if we have `let value: Box<T> = Box::new(T);`, we can call methods on `value` as though it were a `T` without dereferencing it, as in `value.method()`.

Chapter 7 discusses wrapper structs and shows how using the `Deref` trait allows us to treat a wrapper struct as though it were the type it wraps. This situation is a common use case for `Deref`, similar to Rust's smart pointers, which also use `Deref` this way, but it's not polymorphism. In many cases, you can avoid using `Deref` to emulate polymorphism by using traits and generics or by simply providing a method that returns the inner value as required. The following listing illustrates the use of `Deref` in a simple example that uses `Deref` coercion to return the first member of the tuple struct `Person`.

Listing 10.9   Demonstrating `Deref` coercion

```
use std::ops::Deref;

struct Person(String, String, u32);          A tuple struct with a first
                                              name, last name, and age

impl Deref for Person {
    type Target = String;                     Implements the Deref trait for
                                              Person to allow dereferencing
                                              into a String
    fn deref(&self) -> &Self::Target {
        &self.0                               Implements the deref() method
    }                                         to return a reference to the
}                                             wrapped String

fn main() {
    let ferris = Person("Ferris".to_string(), "Bueller".to_string(), 17);
    println!("Hello, {}!", *ferris);
    println!("The length of a person is {}", ferris.len());
}

Dereferences name to                                    Calls the len() method on name
get the inner String                                    as though it were a String via
                                                        Deref coercion
```

In this example, we have a tuple struct, `Person`, that wraps two strings (first and last name) and an age. We can call the `len()` method on `ferris` as though it were a

String thanks to Deref coercion. In this example, we're returning the person's first name, but it's not immediately apparent to the reader why we do this. Why not return the first name directly from a method? Someone who's looking at this code would be confused because it's not idiomatic Rust. We're making Person behave like a String for a particular use case, but why we would want to do so is unclear. When we run the preceding code, it produces the following output:

```
Hello, Ferris!
The length of a person is 6
```

We could just as easily have implemented a first_name() method that returns the inner String or even provided a first_name_len() method, which would be much clearer (though if we return the string, that would be sufficient for getting the length with ferris.first_name().len()). The small convenience of accessing the first name isn't worth the ambiguity introduced by Deref. If we want to provide a first_name_len() method, we could implement it as follows:

```
impl Person {
    fn first_name_len(&self) -> usize {
        self.0.len()
    }
}
```

To see what using Deref to emulate polymorphism looks like, see the following listing, which provides one way to emulate polymorphism in Rust. The example shows a Dog that implements the Animal trait and a Cat that also implements the Animal trait.

Listing 10.10   Emulating polymorphism with trait objects (good practice)

```
trait Animal {
    fn speak(&self) -> &str;
    fn name(&self) -> &str;
}

struct Dog {
    name: String,
}
impl Dog {
    fn new(name: &str) -> Self {
        Self {
            name: name.to_string(),
        }
    }
}
impl Animal for Dog {
    fn speak(&self) -> &str {
        "Woof!"
    }
    fn name(&self) -> &str {
        &self.name
```

```
        }
    }

    struct Cat {
        name: String,
    }
    impl Cat {
        fn new(name: &str) -> Self {
            Self {
                name: name.to_string(),
            }
        }
    }
    impl Animal for Cat {
        fn speak(&self) -> &str {
            "Meow!"
        }
        fn name(&self) -> &str {
            &self.name
        }
    }
```

We can test the code in listing 10.10 by running the code in the following listing. Let's create a vector of Box<dyn Animal> and call the speak() method on each Animal in the vector.

Listing 10.11   Testing the polymorphism with trait objects

```
fn main() {
    let dog = Box::new(Dog::new("Rusty"));
    let cat = Box::new(Cat::new("Misty"));

    let animals: Vec<Box<dyn Animal>> = vec![dog, cat];

    for animal in animals {
        println!("{} says {}", animal.name(), animal.speak());
    }
}
```

Running this code produces the following output:

```
Rusty says Woof!
Misty says Meow!
```

The example is idiomatic Rust. We've used a trait object to create a vector of speaking animals.

Let's create something similar but use Deref to emulate polymorphism this time. We'll create an Animal struct with a name property and treat it as a superclass of the Dog and Cat structs by returning the inner Animal with a Deref.

```rust
use std::ops::Deref;

struct Animal {
    name: String,
}
impl Animal {
    fn new(name: &str) -> Animal {
        Animal { name: name.to_string() }
    }
    fn name(&self) -> &str {
        &self.name
    }
}

struct Dog(Animal);
impl Dog {
    fn new(name: &str) -> Self {
        Self(Animal::new(name))
    }
    fn speak(&self) -> &str {
        "Woof!"
    }
}
impl Deref for Dog {
    type Target = Animal;
    fn deref(&self) -> &Self::Target {
        &self.0
    }
}

struct Cat(Animal);
impl Cat {
    fn new(name: &str) -> Self {
        Self(Animal::new(name))
    }
    fn speak(&self) -> &str {
        "Meow!"
    }
}
impl Deref for Cat {
    type Target = Animal;
    fn deref(&self) -> &Self::Target {
        &self.0
    }
}
```

We can test the code in listing 10.12 by running the code in the following listing.

```rust
fn main() {
    let dog = Dog::new("Rusty");
    let cat = Cat::new("Misty");
```

```
    println!("{} says: {}", dog.name(), dog.speak());
    println!("{} says: {}", cat.name(), cat.speak());
}
```

Running this code produces the following output:

```
Rusty says: Woof!
Misty says: Meow!
```

This example isn't idiomatic Rust. We've used `Deref` to emulate polymorphism in a way that's confusing to anyone who reads the code. The `Animal` struct isn't a super-class of `Dog` and `Cat`, and it's unclear why we would want to treat `Dog` and `Cat` as `Animal`.

You shouldn't avoid `Deref` entirely, but you should avoid overusing or misusing it, especially in contexts that may be confusing or misleading. If you find yourself using `Deref` because you want to emulate the polymorphism of Java or C++, it's probably a good idea to take a step back and think about your design.

10.7 Global data and singletons

Rust doesn't have a built-in concept of global data or singletons, and implementing these concepts requires some work. This is by design, as global data and singletons are often considered antipatterns in programming and can lead to a variety of problems, such as tight coupling, poor testability, and difficulty in reasoning about code.

In Rust, you can use crates like `lazy_static` to create global data or singletons, but you should always think twice before doing so. In many cases, you can avoid global data or singletons by using dependency injection or passing data around as arguments to functions.

For libraries in particular, global data and singletons can be problematic: they make it difficult to reason about the behavior of the library, and they can lead to unexpected behavior when the library is used in different contexts. Global data can become a bottleneck or a source of deadlocks in multithreaded programs, and it can lead to memory leaks and other resource management problems.

Rust provides `std::cell::OnceCell` and its thread-safe counterpart `std::sync::OnceLock`, which give you a safe way to create singletons. As an alternative, your library can provide a function that initializes the singleton, and you can decide how to manage the singleton. This approach is a good way to provide flexibility and avoid the problems associated with global data and singletons.

10.8 Too many smart pointers

Smart pointers are incredibly useful, and in Rust specifically, they're necessary for doing many things that are trivial to do in other languages. It's possible to overuse smart pointers, however, and it's also possible to use the wrong smart pointer for the job. Rust provides the following core smart-pointer types:

- Box—A smart pointer that provides heap allocation and deallocation and allows you to move values between scopes. Box also enables you to hold values whose size isn't known at compile time within objects that have a fixed size (such as Sized).
- Rc—A reference-counted smart pointer that allows multiple owners or shared ownership of a value. It also provides the features of Box.
- Arc—An atomic reference-counted smart pointer that allows multiple owners of a value across threads, providing the features of Rc and Box in a thread-safe manner. Arc doesn't synchronize the value itself; its synchronization is only for the reference count.

Generally, you use Box when you need heap-allocated memory but don't need to share ownership of the value. You use Rc when you need to share ownership of a value but don't need to share ownership across threads. You use Arc when you need to share ownership of a value across threads.

Additionally, RefCell and Cell provide interior mutability and are often used in conjunction with Rc and Arc. Rc and Arc allow you to share ownership of a value but don't allow you to mutate the value. RefCell and Cell allow you to mutate the value but don't allow you to share ownership of the value.

Sometimes, however, we reach for smart pointers when we don't need them or overuse them simply because they're convenient or allow us to get around the borrow checker. It can be easier to stick something into Rc and clone the pointer than it is to think about the ownership and borrowing of the value.

Another example of overusing smart pointers is using Box within a Vec. Both Box and Vec allocate memory on the heap for their contents, so this approach can be redundant and requires two allocations: one for Vec and another for the contained Box. The following listing shows an example of triple allocation: String within Box within Vec.

Listing 10.14   Overusing smart pointers

```
let mut string_box_vec: Vec<Box<String>> =
    vec![Box::new(                                          Redundantly boxes a
        String::from("unecessarily boxed string")          String within a Box
    )];                                                     within a Vec
let mut string_vec: Vec<String> =
    vec![String::from("this is okay")];         ←─────      Stores a String
                                                            directly within a Vec

let boxed_string = string_box_vec.remove(0);   ←─┐         Removes the boxed String from
let normal_string = string_vec.remove(0);      ←─┐         the Vec, which is equivalent to
                                                          removing a pointer to a pointer
        Removes the String from the Vec, which is         in the Vec
            equivalent to removing a single pointer
```

A String is equivalent to a pointer with length to the heap, and a Box is a pointer to some heap-allocated value. We sometimes reach for Box because it allows us to move values between scopes, but in this case, doing so is redundant because Vec already provides this functionality. It's okay to put a `String` inside a `Vec` because strings are variable-length, and each entity within a `Vec` requires a fixed (and equal) size.

If you use smart pointers as an escape hatch to avoid the borrow checker, you should reconsider your design. A good rule of thumb is to try to write code without smart pointers and add them as required.

10.9 Where to go from here

When you finish reading this book, the most critical steps for leveling up your skills are writing code and applying what you've learned. Practice is the best way to learn, and you'll learn a lot by writing code, getting feedback on your code, and reading other people's code. You may want to refer to this book as you progress in your learning, and you may discover that you learn even more after some practice. If you want to read more books on Rust, you may be interested in my book *Code Like a Pro in Rust* (https://www.manning.com/books/code-like-a-pro-in-rust), which inspired this book.

The official Rust documentation is an excellent resource, and familiarizing yourself with it is a good idea. Also, the Rust community is very active and offers many resources, including the Rust subreddit at https://reddit.com/r/rust, the Rust Discord server at https://discord.gg/rust-lang, and the Rust user forums at https://users.rust-lang.org.

Finally, many Rust meetups and conferences enable you to get involved in the community and meet other Rustaceans in person. Sometimes, you can find me at the New York City Rust meetup; I'm always happy to chat about Rust and programming in general.

> **TIP** At the NYC meetup, we have a tradition of answering questions from David Tolnay's Rust Quiz, which is a fun way to hone skills and learn about some of Rust's more esoteric syntax and features. You can find the quiz at https://dtolnay.github.io/rust-quiz.

Summary

- Antipatterns are programming practices that are considered harmful, either in specific contexts or all circumstances. Although the use of antipatterns is often a matter of opinion, in some cases, it's objectively bad, such as when it's unsafe, inefficient, or difficult to maintain.
- The `unsafe` keyword is a necessary part of Rust that is sometimes misused or overused. It's nearly impossible to use Rust without using `unsafe` code (at least indirectly), but you should scrutinize its use when you come across it. You should never use `unsafe` to bypass the borrow checker.
- The `unwrap()` method is a common antipattern in Rust, often used when we get lazy about handling `Option` or `Result` values. It's relatively easy to avoid `unwrap()` by replacing it with one or more of the following methods: `expect()`, `map()`, `and_then()`, `unwrap_or()`, and the `?` operator.

- Vec is fast for many workloads and is often the best choice. It's often faster across a variety of benchmarks than HashSet, HashMap, BTreeSet, BTreeMap, and LinkedList, and it's also more memory-efficient.

- The clone() method is overused sometimes and often used when it's not necessary. It's not always bad but can be a code smell, leading to performance problems and memory bloat.

- The Deref trait is sometimes used to emulate polymorphism, which can be confusing in Rust. Instead, you should rely on traits or generics or simply provide a method that returns the inner value as required.

- Global data and singletons are often considered antipatterns in programming. They can lead to a variety of problems, such as tight coupling, poor testability, and difficulty in reasoning about code. In Rust, you can use crates such as lazy_static to create global data or singletons, but always think twice before doing so.

- Smart pointers are incredibly useful, but it's possible to overuse them or use the wrong smart pointer for the job. If you use smart pointers as an escape hatch to avoid the borrow checker, think about your design.

appendix
Installing Rust

To get the most out of this book, you'll need to have a functioning Rust toolchain installed. If you've never used Rust before, you'll need to install a recent release of the Rust toolchain that includes the compiler and the standard library. You may also need to install some development tools, depending on your OS, to compile and run all the code samples included with this book.

A.1   Installing tools for this book

To compile and run the code samples provided in this book, you must install the necessary prerequisite dependencies.

A.1.1   Installing tools for macOS using Homebrew

```
$ brew install git
```

In macOS, you'll need to install the Xcode command-line tools:

```
$ sudo xcode-select --install
```

A.1.2   Installing tools for Linux systems

To install tools for Debian-based systems, use this command:

```
$ apt-get install git build-essential
```

To install tools for Red Hat-based systems, use this command:

```
$ yum install git make automake gcc gcc-c++
```

> **TIP** You may want to install Clang rather than GCC, which tends to have better compile times.

To install `rustup` in Linux and UNIX-based operating systems, including macOS, use this command:

```
$ curl --proto '=https' --tlsv1.2 -sSf https://sh.rustup.rs | sh
```

When you've installed `rustup`, make sure that both the stable and nightly toolchains are installed:

```
$ rustup toolchain install stable nightly
...
```

A.1.3 Installing tools for Windows

If you're using a Windows-based OS, you'll need to download the latest copy of `rustup` at https://rustup.rs. You can download prebuilt Windows binaries for Clang at https://releases.llvm.org/download.html.

Alternatively, you can use Windows Subsystem for Linux (WSL; https://docs.microsoft.com/en-us/windows/wsl) and follow the instructions in the preceding section for installation in Linux. For many users, this approach may be the easiest way to work with the code samples.

A.2 Managing rustc and other Rust components with rustup

When you have `rustup` installed, you'll need to install the Rust compiler and related tools. At a minimum, I recommend that you install the stable and nightly channels of Rust.

A.2.1 Installing rustc and other components

By default, you should install both the `stable` and `nightly` toolchains, but generally, you should prefer working with `stable` when possible. To install both toolchains, use this code:

```
# Install stable Rust and make it the default toolchain
$ rustup default stable
...
# Install nightly Rust
$ rustup toolchain install nightly
```

Examples throughout this book use `clippy` and `rustfmt`, both of which you install by using `rustup`:

```
$ rustup component add clippy rustfmt
```

A.2.2 Switching default toolchains with rustup

When working with Rust, you may switch between the stable and nightly toolchains frequently. `rustup` makes this switch relatively easy:

```
# Switch to stable toolchain
$ rustup default stable
# Switch to nightly toolchain
$ rustup default nightly
```

A.2.3 *Updating Rust components*

`rustup` makes it easy to keep components up to date. To update all the installed toolchains and components, simply run

```
$ rustup update
```

Under normal circumstances, you need to run `update` only when major new releases are available. Occasionally, problems in `nightly` require an update, but they tend to be infrequent. If your installation is working, it's recommended you avoid upgrading too frequently (i.e., daily) because you're more likely to run into problems.

> **NOTE** Updating all Rust components causes all toolchains and components to be downloaded and updated, which may take some time on bandwidth limited systems.

index